Microbial Ecology

Current Advances from Genomics, Metagenomics and Other Omics

https://doi.org/10.21775/9781912530021

Edited by Diana E. Marco

Faculty of Biological Sciences
Córdoba National University
Argentina

and

CONICET
Córdoba
Argentina

Caister Academic Press

Caister Academic Press
Norfolk, UK

www.caister.com

British Library Cataloguing-in-Publication Data
A catalogue record for this book is available from the British Library

ISBN: 978-1-912530-02-1 (hardback)
ISBN: 978-1-912530-03-8 (ebook)

Cover design adapted from Figure 4.5.

Contents

Current Books of Interest

Methylotrophs and Methylotroph Communities	2019
Plant-Microbe Interactions in the Rhizosphere	2018
Prions: Current Progress in Advanced Research (Second Edition)	2019
Microbiota: Current Research and Emerging Trends	2019
Lactobacillus Genomics and Metabolic Engineering	2019
Cyanobacteria: Signaling and Regulation Systems	2018
Viruses of Microorganisms	2018
Protozoan Parasitism: From Omics to Prevention and Control	2018
Genes, Genetics and Transgenics for Virus Resistance in Plants	2018
DNA Tumour Viruses: Virology, Pathogenesis and Vaccines	2018
Pathogenic *Escherichia coli*: Evolution, Omics, Detection and Control	2018
Postgraduate Handbook: A Comprehensive Guide for PhD and Master's Students and their Supervisors	2018
Enteroviruses: Omics, Molecular Biology, and Control	2018
Molecular Biology of Kinetoplastid Parasites	2018
Bacterial Evasion of the Host Immune System	2017
Illustrated Dictionary of Parasitology in the Post-Genomic Era	2017
Next-generation Sequencing and Bioinformatics for Plant Science	2017
The CRISPR/Cas System: Emerging Technology and Application	2017
Brewing Microbiology: Current Research, Omics and Microbial Ecology	2017
Metagenomics: Current Advances and Emerging Concepts	2017
Bacillus: Cellular and Molecular Biology (Third edition)	2017
Cyanobacteria: Omics and Manipulation	2017
Brain-eating Amoebae: Biology and Pathogenesis of Naegleria fowleri	2016
Foot-and-Mouth Disease Virus: Current Research and Emerging Trends	2017
Staphylococcus: Genetics and Physiology	2016
Chloroplasts: Current Research and Future Trends	2016
Microbial Biodegradation: From Omics to Function and Application	2016
Influenza: Current Research	2016
MALDI-TOF Mass Spectrometry in Microbiology	2016
Aspergillus and *Penicillium* in the Post-genomic Era	2016

Full details at www.caister.com

Preface

The development of 'meta-omics' methods such as metagenomics, metatranscriptomics, metaproteomics, metametabolomics and other related methods is greatly contributing to the understanding of the complexity of interactions among microorganisms and of the interaction of microorganisms with their environment and with other organisms.

As in any ecological framework, the concepts of pattern, process and mechanism are of primary importance in the microbial ecology context and there is a pressing need to investigate the processes and mechanisms that may explain the occurrence of detected patterns. Some operative definitions and considerations to help microbial ecology researchers to become acquainted with these ecological concepts are presented in Chapter 1. This chapter also provides an account of the contributions of 'meta-omics' methods to the identification and understanding of patterns, processes and mechanisms in microbial ecology. The 'omics' approaches are also advancing the field of microbial ecology by developing new molecular methods to study microbiomes. As contamination in microbiome sequencing studies is a well-known but difficult to address problem, Chapter 2 presents a simulation model to examine when contamination is likely to be problematic in microbiome samples. Chapter 3 provides an overview of the advances in molecular methods based on DNA sequencing that improve our understanding of the dynamics of microbial communities in soil systems, in particular the study of patterns of microbial succession in soils. Chapter 4 demonstrates, by the development of an insular microbiogeographical model, that the same mechanism that underlies macro-ecological scaling laws also applies to microbial communities. This finding is useful in understanding the diversity and dynamic exchange of genes. Chapter 5 explores the contribution of metagenomics in the understanding of complex sediment microbial communities, an issue that remains a challenge because of the remarkable diversity of these communities. Finally, Chapter 6 addresses, in a novel interdisciplinary way, how genomic approaches to microbial ecology can be combined with host biology, endocrinology and disease data to characterize the overall health of animals, habitats and ecosystems. This new avenue is interesting from a theoretical point of view but also as a potential method to improve conservation and disease surveillance strategies.

The book is aimed at scientific researchers, educators and advanced students interested in approaching the microbial ecology field by utilizing the most recent and advanced 'omics' methods. The book covers both the theoretical and the applied aspects of microbial ecology.

I would like to thank all of the authors for their contributions. I am also very grateful to Hugh Griffin and Caister Academic Press for giving me the opportunity of editing a volume on such an interesting and dynamic field.

Diana Marco
Granada, Spain

Patterns, Processes and Mechanisms in Microbial Ecology: Contributions from the 'Omics'

1

Diana E. Marco[1,2*]

[1]Faculty of Biological Sciences, Córdoba National University, Argentina.
[2]CONICET, Córdoba, Argentina.

*Correspondence: dmarco@agro.unc.edu.ar

https://doi.org/10.21775/9781912530021.01

Abstract

The complexity of interactions among microorganisms and of microorganisms with their environment and with other organisms is becoming increasingly understood by microbial ecology researchers, helped by the development of 'meta-omics' methods such as metagenomics, metatranscriptomics, metaproteomics, metametabolomics and other related techniques.

As in any ecological framework, the concepts of pattern, process and mechanism are of primary importance to the microbial ecology context. The observation and description of ecological patterns are not enough and there is a need to seek the processes and mechanisms that may explain the occurrence of these patterns. However, as there is some confusion about the meaning and importance of these concepts in microbial ecology studies, some operative definitions and considerations to help microbial ecology researchers to become acquainted with these ecological concepts are provided in this chapter. A brief account of the contributions from the 'meta-omics' to the identification and understanding of patterns, processes and mechanisms in microbial ecology is also provided.

This chapter is focused on the community and biogeographical levels of soil microbiomes studied using some of the 'meta-omics' methods, as much of the recent work has been performed at these ecological levels and in this habitat.

Introduction

> Understanding patterns in terms of the processes that produce them is the essence of science . . . Without an understanding of mechanisms, one must evaluate each new stress on each new system *de novo*, without any scientific basis for extrapolation.
>
> Simon A. Levin (1992)

The field of microbial ecology has undergone an exponential growth in recent years because of the advent of new techniques allowing for a better comprehension of the structure and function of an increasing variety of microbiomes. From the limitations of the methods based on isolation and cultivation of a particular microorganism (although still valid for specific questions) to understanding complex microbiomes, researchers have begun in recent years to study the composition and function of entire microbial communities using 'meta-omics' methods such as metagenomics, metatranscriptomics, metaproteomics, metametabolomics and other related techniques. The complexity of interactions among microorganisms and of the interaction of microorganisms with their environment and with other organisms is being increasingly understood by microbial ecology researchers. However, many studies are not question or hypothesis driven, and microbiomics is still dominated by descriptive and correlation-based studies (Prosser, 2017). Central to this is the fact that, as in any other ecological field, the concepts of pattern, process and mechanism in microbial ecology ought to be clearly understood and taken into account to formulate suitable hypotheses to be tested with the appropriate techniques. As pointed out by Simon Levin in his MacArthur Award Lecture (Levin, 1992), observation and description of ecological patterns are not enough and there is the need to seek the processes and mechanisms that may explain the occurrence of those patterns. At the same time, it should be kept in mind that patterns, processes and mechanisms are scale dependent and that to adequately detect and understand them it is essential to define the temporal and spatial scales of the study that will allow for relevant hypotheses formulation. The scale issue is of primary importance in microbial ecology as composition, structure and functional traits of microbiomes are characterized by a high heterogeneity both in space and in time, a property that influences even the sampling design in field microbial ecological studies (Marco, 2017).

This chapter intends to bring the ecological concepts of pattern, process and mechanism into the microbial ecology context, highlighting the importance of the temporal and spatial scales. The chapter is focused on the community and biogeographical levels of soil microbiomes studied using some of the 'meta-omics' methods as much of the recent work has been performed at these ecological levels and in this habitat.

Ecological definitions of pattern, process and mechanism

Although the concepts of pattern, process and mechanism are amply used in ecology, it is difficult to find precise definitions of them in the literature. Moreover, the terms 'process' and 'mechanism' are sometimes used in an interchangeable way. To avoid confusion, here I give some operative definitions that may help to clarify this issue.

A 'pattern' is a set of data points of two or more variables such that the values of the variables show some relationship to each other. This relationship may range from simple linearity to complex nonlinear chaos (Rosenzweig, 1999). In addition, a random relationship may originate 'random patterns'. A 'process' is the dynamic change of different variables in an ecological setting. A 'mechanism' is the 'ultimate' cause of such changes underlying a process. The level of the 'ultimate' cause (mechanism) explained will depend on the knowledge of a given system. By working together at different levels, process and mechanism lead to the detectable pattern (Fig. 1.1).

A simplified example will clarify the definitions. A patchy vegetation spatial pattern (detectable through aerial images) may be explained by differential plant biomass growth (process) caused ultimately by a particular spatial distribution of a soil nutrient (mechanism). Translating the example to microbial ecology, finding a patchy distribution of a soil microbial community of denitrifying bacteria could be explained as differential bacterial biomass growth in response to an also patchy distribution of nitrate in the soil (Correa-Galeote *et al.*, 2013). However, it is important to note that a given pattern could be explained by alternative processes and mechanisms, although this arrangement of hierarchical 'causalities' may sound attractively simple (Fierer, 2017). To determine the actual processes and mechanisms acting behind an observed ecological pattern, the appropriate field or

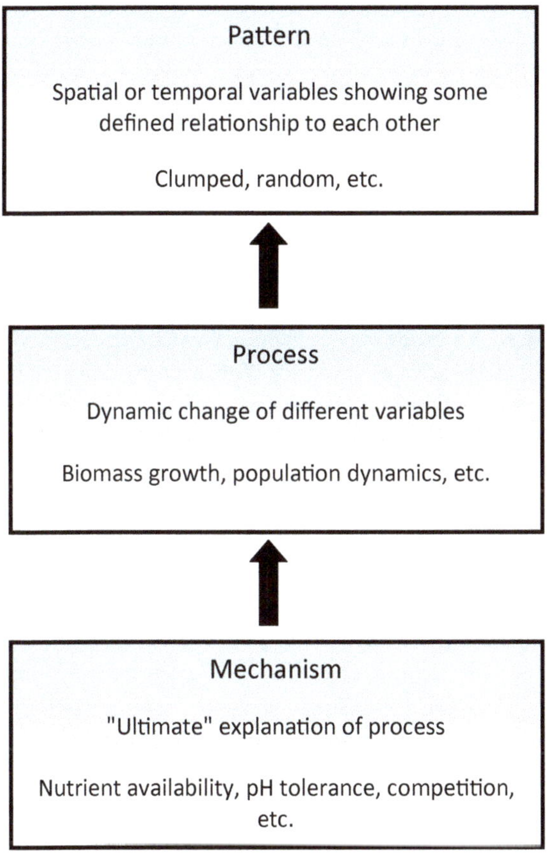

Figure 1.1 Schematic representation of the ecological concepts of pattern, process and mechanisms and their relationship.

laboratory (or both) sampling and experiments should be devised, formulating the right hypotheses/questions, with the appropriate replicates and taking into account the correct spatial and temporal scales (Hurlbert, 1984; Prosser, 2010; Marco, 2017). Although these requirements may sound quite obvious, many studies conducted nowadays in microbial ecology still disregard some or all of these requirements (Prosser, 2015, 2017).

The problems of scale, heterogeneity and replication

The relevance of spatial and temporal scaling and the environmental heterogeneity issues, along with their implications for an adequate sampling and/or experimental design in microbial ecology, have been recently approached (Marco, 2017). However, here I will briefly refer to some concepts useful to a further understanding of how the new 'omics' methods are improving the detection of patterns and the comprehension of the processes and mechanisms behind them.

Choosing a particular spatial scale for sampling may be decisive for the kind of patterns found (or to find any pattern) (Wiens, 1989). Central to the scaling problem are the concepts of *extent* and the *grain* of a study (O'Neill *et al.*, 1986). The whole area encompassed by a study to be described by sampling is the extent of the study, whereas the grain is the size of the individual units of observation. The homogeneity or heterogeneity of the extent considered, and the grain size of the study, will condition finding a pattern. Increasing the grain size means that a greater proportion of the spatial heterogeneity of the system is contained within a sample and that part of this heterogeneity is lost to the study resolution (Wiens, 1989). The extent and grain of a study should be defined by our knowledge of the system, to help us to discern, for example, the effects of physical processes that could act at broader scales from more local, edaphic or biological interactions acting as mechanisms, and should depend on the hypothesis and aim of the study. The 'domains of scale' are regions of the scale spectrum over which patterns either do not change or change monotonically, with changes in scale for a given phenomenon in a particular ecological system. Relatively sharp transitions separate domains from changes in dominance by one set of factors to dominance by other sets (Wiens, 1989). Since the early 20th century different methods have been used in ecology to assess spatial heterogeneity and to detect scale domains (summarized in Marco, 2017).

In close relationship with the spatial heterogeneity of many ecological systems in nature, there is the problem of spatial pseudoreplication in field and experimental studies. Pseudoreplication is the use of inferential statistics to test for treatment effects with data from experiments where either treatments are not replicated (even when samples may be) or replicates are not statistically independent (Hurlbert, 1984). Pseudoreplication is an extended problem in microbial ecology studies even now (Prosser, 2010), because either 'simple pseudoreplication' (no true replicates) or 'sacrificial pseudoreplication' (pooling samples from true replicates) (Hurlbert, 1984) is still commonly performed in field studies.

The same rules applying for a classical ecological study, guiding the extent and grain selection and statistical replication, should be followed when formulating the hypothesis and designing a spatial sampling or experimental design for a microbial ecology study. Soil microbiomes are highly spatially heterogeneous (Cordero and Datta, 2016), so choosing the adequate spatial scale for a study is fundamental. Different drivers (processes and mechanisms) of microbiome community composition are expected to act at each length scale. In soil, the main drivers at the ecosystem (regional and biogeographic) scales (> m) are factors like climate and biogeochemical processes, whereas environmental gradients (pH, soil moisture, etc.) are the main drivers at meta-community scales (cm to m). At the microbial community level (10–10^3 μm) local ecological interactions govern the structure and functioning of soil microbial aggregations (Cordero and Datta, 2016). Local interactions arise mainly because the soil is extremely heterogeneous and patchy, composed of micro-aggregates (at the 10 μm scale), with micropores water-filled, and further clustered into macro-aggregates with meso- and macropores (at the 100 μm scale) filled with water or air (Vos *et al.*, 2013). The patchy distribution of resources and incomplete connectivity restrict nutrient access by microbial cells and the interaction with other cells, thus creating very local microbial heterogeneity that can drive ecological and evolutionary processes (Vos *et al.*, 2013; Rillig *et al.*, 2017). Other factors contribute to microbial patchiness, like cell division (resulting in a short distance dispersal) and the microbial ability of creating biofilms. Biofilm (structured aggregations of one or more kinds of microbial cells) production and activity may also create heterogeneity from homogeneous initial conditions (Nadell *et al.*, 2016). This doubly originated heterogeneity (from the soil and from the microbes) makes grain size selection difficult in microbial ecology studies and may explain the failure of many studies to detect microbial patterns and significant correlations between microbial community variables and habitat variables like soil pH, moisture and other factors postulated as potential drivers (processes and mechanisms) of soil microbiome composition and distribution (Marco, 2017). For example, addressing the *pmoA* genes as phylogenetic markers of aerobic methane-oxidizing bacteria in paddy soils at the 100-μm spatial scale, Reim *et al.* (2012) found that different operational taxonomic units (OTUs) within a single guild shared the same microenvironment but exploited different niches. If the authors had used a broader spatial scale, detection of this community structure could have been missed.

In ecology, temporal scales are inherently connected with spatial scales. Increasing the spatial scale, the timescale of relevant processes also increases because they operate at slower rates, time lags increase and indirect effects gain importance (Wiens, 1989). Therefore, it is not surprising that there is a tendency to approach simultaneously both scales in ecological studies (Legendre and Gauthier, 2014). In soil microbiomes, changes in community composition and functioning may occur over seasonal temporal scales. For example, in alpine soils, there is an almost complete turnover of the microbial community composition comparing winter and summer (Schadt *et al.*, 2003) and also functional attributes differ (Lipson and Schmidt, 2004). Taking one sampling in a given time of the year will give a mistaken idea of the microbiome composition and functioning in this kind of soils. Because of processes and mechanisms like predation of microbes by bacteriophages, soil protozoa, the action of abiotic stresses and temporal variation in the supply of carbon and other nutrients from roots to soil, shorter timescales occurring over hours to days are also important in shaping soil microbiome composition and functioning (Mikola *et al.*, 2002; Bardgett *et al.*, 2005). Thus, the problem of pseudoreplication in temporal microbial ecology studies is as important as in spatial studies. A variety of techniques is increasingly being used to address the problem of temporal sampling in microbial ecology studies (reviewed in Marco, 2017), which can help to distinguish stochastic fluctuations from significant temporal patterns (Hekstra *et al.*, 2012).

Finally, it should be stressed that the choice of spatial and temporal scales of a study will determine the patterns, processes and mechanisms that can be detected and understood. As already mentioned, by studying a system at an inappropriate scale, its actual dynamics and patterns may be not detected but instead a pattern that is merely an artefact of scale may be found or no pattern may be detected at all. To overcome these problems, a useful ecological approach, multiscale analysis, is increasingly used. The method consists of performing an analysis with respect to multiples of a unit of measurement (Schneider, 1994). By changing the grain of the study (the sampling quadrat size in a given extent in a spatial study, for example), the values of diversity indexes or other microbiome community variables are expected to change. This approach allows, for example, to detect changes in the scale domains of variables of soil microbiomes. However, it should be noted that this is not the same as only changing the extent of the study, that is, simply spanning many quadrats of equal size in a wider space. The multiscale method ought to be used to assess changes in time as well. Nevertheless, studies with a temporal multiscale approach are only beginning to be conducted in microbial ecology (Stempfhuber, 2016). Unified spatial and temporal designs for microbial ecology studies together with multiscale approaches (including environmental variables as well) will allow us to detect and explain the high variability of microbial communities (Gonzalez *et al.*, 2012; Gilbert and Henry, 2015; O'Brien *et al.*, 2016).

Contributions from the 'meta-omics' to the identification and understanding of patterns, processes and mechanisms in microbial ecology

The recently developed 'meta-omics' approaches, increasingly used in microbial ecology, including metagenomics, metatranscriptomics, metaproteomics, metametabolomics, lipidomics and emerging approaches like DNA stable isotope probing (SIP), single-cell analysis and other methods (Meiring, 2011), are making decisive contributions to the discovery and understanding of microbiome ecological patterns, processes and mechanisms. They are also greatly contributing to the two main questions in microbial ecology, 'Who is there'? and 'What are they doing'?, involved in the 'open' (a particular sequencing is not addressed beforehand) and 'closed' (a particular gene or function is targeted) methodological approaches, leading to a deeper understanding of microbiomes and new functional discoveries (Zhou *et al.*, 2015).

This section intends to give a brief account of the significant contributions of 'meta-omics' methods already made and those expected in the near future to microbial ecology patterns, processes and mechanisms detection and understanding. Because of space limits, this section is focused on soil microbial diversity at community and biogeographical levels.

Soil microbial diversity

The first conclusion that can be drawn from the literature when looking for a consistency in the patterns, processes and mechanisms governing the soil microbiome community structure (composition and diversity), is that not a single biotic or biotic factor may be invoked as explaining the found patterns, at biogeographical and local levels (Fierer, 2017). This is perhaps an expected consequence of the high heterogeneity of soil habitat conditions described above. However, a consensus is emerging regarding the variables reported as most likely to have marked effects on the composition and diversity of soil microbiomes (Fierer, 2017; Thompson *et al.*, 2017). Based on studies examining spatial patterns in soil microbial communities, Fierer (2017) constructed a hierarchical arrangement of abiotic and biotic factors influencing soil bacterial communities across space or time, indicating their relative importance. The factors were also classified following their degree of understanding by microbial ecology researchers. Soil variables like pH, carbon content and redox status were the more relevant and also the most studied. This is followed by a set of four soil variables, of gradually lower importance and decreasing understanding, such as soil moisture, nitrogen and phosphorus availability, soil texture and structure, and soil temperature. Finally, there are two variables more related to biotic factors, of much lower importance and less studied: plant species occupying the soil, and predation and viral lysis. However, although this arrangement of factors may serve as a preliminary guide, it does not consider differences in the spatial and temporal scales of the studies used to construct it, and also makes no distinction between factors working as processes or mechanisms at a given scale.

Even when finding global patterns in microbiome diversity is difficult (Fierer, 2017), the use of 'meta-omics' methods is increasingly facilitating the detection of patterns. However, the identification and understanding of processes and mechanisms that could explain these patterns are still difficult. For example, by exploring an impressive amount of studies of different habitats at the global scale from the Earth Microbiome Project (EMP, http://www.earthmicrobiome.org) in a metadata analysis, Thompson *et al.* (2017) found that a maximum microbial richness occurs within a relatively narrow range of intermediate pH and temperature values in soil. However, as the authors themselves warn, these patterns, although consistent within the EMP database, should be interpreted with caution as they came from a limited subset of sample types and metadata variables. Moreover, although the EMP promotes the use of standardized protocols for DNA extraction and sequence processing, there is a disparity of spatial and timescales sampling among the different studies. Thus, the difficulties of finding and understanding patterns, processes and mechanisms of soil microbial diversity arise most probably not from the suitability of the 'meta-omics' method themselves, but from issues concerning the scales (spatial and temporal) and sampling replication. Some relevant examples will illustrate this.

The most common 'meta-omics' methods used to study soil microbiome composition and diversity are sequencing-based, open-format methods involving DNA/RNA extraction from soil samples prepared for high-throughput sequencing, for example shotgun metagenomics, and/or sequencing by target gene sequencing (TGS) (Zhou *et al.*, 2015). This approach has allowed non-random biogeographical patterns of soil microbial diversity at continental scale to be found. For example, Barberán *et al.* (2012) collected 151 soil samples across North and South America and Antarctica, from a variety of ecosystems, climates and soil types, and performed 16S rRNA gene pyrosequencing to investigate potential biogeographical patterns in microbial communities. A network analysis of significant taxon co-occurrence patterns was afterwards performed, revealing common life history strategies at broad taxonomic levels and unexpected relationships between community members (Barberán *et al.*, 2012). However, although some insights could be extracted to characterize the found patterns, like the distribution of habitat 'generalist' and 'specialist' bacteria and archaea, no processes and mechanisms were explicitly investigated. Given the heterogeneity of sampling sites and the sampling method used in this study (collecting several soil samples in each location inside a quadrat and then pooling them together in a single sample), it is difficult to assert the actual spatial scale of the patterns found, and even more difficult to decide the adequate scale level to detect processes and mechanisms.

In another example at intercontinental scale, Shi *et al.* (2015), looked for biogeographical patterns of microbial functional genes from 24 heath soils across the Arctic using GeoChip metagenomics, a kind of method classified as 'closed' (Zhou, 2015). Sampling sites were scattered over the Canadian, Alaskan and European Arctic in a broad extent and the grain used was the same through the locations (sampling quadrats of 12 × 12 cm). The authors found that 20% of the variation in total and major functional gene categories could be attributed primarily to relatively large-scale spatial effects (spatial correlations), consistent with broad-scale variation in pH and total nitrogen of sampled soil. One of the questions formulated by the authors was whether or not the spatial structure of soil microbial functional genes could be categorized into discrete spatial scales that are associated with heterogeneities in environmental variables. The problem in answering this question and assessing processes and mechanisms in this study is that the correlations were measured between quadrats similar in size but at different distances. It is not surprising then that an important outcome of the study was that most of the variance found could be attributed to historical contingencies such as disturbance events or dispersal barriers that occurred in the past, rather than to local environmental variables.

Two aspects are important for going beyond identifying patterns to address the processes and mechanisms behind them. One aspect, already approached in the chapter, is the relevance of setting the right scales (and the appropriate replicated field/experimental designs) of the studies. It is being increasingly accepted that soil microbial diversity may be controlled by processes operating at scales that do not match the temporal and spatial scales commonly used (Gupta and Germida, 2015; Hendershot *et al.*, 2017), and that small-scale processes might be more important than regional processes in determining microbial abundance patterns (Štursová *et al.*, 2016).

The other aspect is the increasing number of methods ('meta-omics' and other) developed or under development to study *in situ* microbial activity and that are amenable to be used in combination with nucleic acid sequencing methods. Depending on the microbial activity targeted (DNA synthesis and transcription and lipid, carbohydrate and amino acid biosynthesis), several techniques have been recently developed (or are under development), which in combination with massive sequencing methods allow microbial cell activities to be studied, as single-cell and/or bulk community analysis (Singer *et al.*, 2017). Among the DNA synthesis methods, DNA stable isotope probing (DNA-SIP) (Warwick, 2014) allows metabolic functions to be linked to identity. DNA-SIP has been used, for example, to investigate soil microbial diversity patterns and the ability to mineralize organic matters of microbial communities as a driving process along climate gradients in Madagascar (Razanamalala *et al.*, 2018). Among the transcriptomics methods, RNA-SIP has been used, although is technically more demanding (Paul *et al.*, 2018). Metatranscriptomics is increasingly being employed, although with the caveat that mRNA abundance may not reflect the actual protein levels and enzymatic activities. Metatrascriptomics has been shown to be especially useful for studying microbiomes in permafrost soils. Because in frozen soil samples bacterial DNA can be preserved for thousands of years (Willerslev *et al.*, 2004), making difficult to distinguish between past and present microbial communities using DNA-based metagenomics, the short-lived mRNA provides information on microbial activities occurring in the permafrost soils at sampling time. Using metatrascriptomics, the microbial communities and their active metabolic pathways were studied in the acidic tundra in Arctic

Alaska, comparing before and after soil thaw events. The transcriptional profiles under frozen conditions mainly showed stress responses, survival strategies and maintenance processes, whereas after thaw a rapid enzymatic response to decomposing soil organic matter was detected (Coolen and Orsi, 2015). Metabolomics has been used in combination with nucleic acid sequencing in desert soil biocrusts (associations of microorganisms inhabiting the top soil stratum) to reveal not only the community structure of microbial entities composing the biocrust but also the mechanisms behind microbial sympatric coexistence, mediated by metabolic niche partitioning (Baran *et al.*, 2015).

Among the techniques based on cell components other than nucleic acids, BONCAT (bioorthogonal non-canonical amino-acid tagging) is being increasingly used. BONCAT is based on the incorporation of artificial amino acids that carry modifiable chemical tags into newly synthesized proteins *in vivo* (Dieterich *et al.*, 2006). This technique has been demonstrated to be effective for labelling the proteomes of a wide range of archaea and bacteria that are taxonomically and physiologically different (Hatzenpichler *et al.*, 2014). BONCAT coupled to high-throughput sequencing will allow the detection of cellular translational activities in response to environmental variables at micrometre as well as across larger scales.

Another technique not based on nucleic acids but allowing instead the chemical fingerprinting of individual cells, including eukaryotic, bacterial and archaeal cells, is Raman microspectroscopy (Huang *et al.*, 2004). Raman can be used in combination with several other methods, like FISH (Huang *et al.*, 2007), and can be used in combination to sort isotope-labelled microbial cells in microcosm experiments (Cui *et al.*, 2018). Further combinations of Raman with other methods are appearing (Huys and Raes, 2018).

This brief account of some of the possible combinations of methods is by no means exhaustive and is continually being updated.

Future trends

The combination and complementary usage of different methods, including 'meta-omics' and others, allows for an increasing understanding of microbial activity *in situ*, which means that microbial ecologists are reaching the ability of not only detecting microbiome patterns but also addressing the processes and mechanisms behind those patterns. As new 'omics' and 'meta-omics' methods are under development, and new combinations of already existing methods (sometimes seen as 'old' and separated from the 'omics' and 'meta-omics') are increasingly being attempted (Paul *et al.*, 2018), the ability of linking microbiome structure and diversity to function will be enhanced. Metagenomics and other sequencing methods are able to account for microbial identities and functional gene information but from microbiomes whose components are in different physiological states, thus revealing just the community functional potential. A new concept, the 'metaphenome', that is, combining genetic information of the microbiome and effective functional activities under different habitat conditions (Jansson and Hofmockel, 2018), seems to be a new avenue for detecting and understanding patterns, processes and mechanisms in microbial ecology.

Valuable initiatives like the creation of a reference database to give global context to nucleic acid sequence data integrating environmental metadata (Thompson *et al.*, 2017) will be very welcome, to help the standardization of sampling and analysing methods and for future studies.

These new approaches will contribute also to the ability to model and predict patterns, processes and mechanisms in microbial ecology, thus allowing for deeper theoretical insights and for potential applications like enzyme and bioproduct discoveries (Medema, 2018).

Acknowledgement

DEM is a research member of the Consejo Nacional de Investigaciones Científicas y Tecnológicas (CONICET), Argentina.

References

Baran, R., Brodie, E.L., Mayberry-Lewis, J., Hummel, E., Da Rocha, U.N., Chakraborty, R., Bowen, B.P., Karaoz, U., Cadillo-Quiroz, H., Garcia-Pichel, F., *et al.* (2015). Exometabolite niche partitioning among sympatric soil bacteria. Nat. Commun. *6*, 8289. https://doi.org/10.1038/ncomms9289

Barberán, A., Bates, S.T., Casamayor, E.O., and Fierer, N. (2012). Using network analysis to explore co-occurrence patterns in soil microbial communities. ISME J. *6*, 343–351. https://doi.org/10.1038/ismej.2011.119

Bardgett, R.D., Bowman, W.D., Kaufmann, R., and Schmidt, S.K. (2005). A temporal approach to linking aboveground and belowground ecology. Trends Ecol. Evol. *20*, 634–641.

Coolen, M.J., and Orsi, W.D. (2015). The transcriptional response of microbial communities in thawing Alaskan permafrost soils. Front. Microbiol. *6*, 197. https://doi.org/10.3389/fmicb.2015.00197

Cordero, O.X., and Datta, M.S. (2016). Microbial interactions and community assembly at microscales. Curr. Opin. Microbiol. *31*, 227–234.

Correa-Galeote, D., Marco, D.E., Tortosa, G., Bru, D., Philippot, L., and Bedmar, E.J. (2013). Spatial distribution of N-cycling microbial communities showed complex patterns in constructed wetland sediments. FEMS Microbiol. Ecol. *83*, 340–351. https://doi.org/10.1111/j.1574-6941.2012.01479.x

Cui, L., Yang, K., Li, H.Z., Zhang, H., Su, J.Q., Paraskevaidi, M., Martin, F.L., Ren, B., and Zhu, Y.G. (2018). Functional single-cell approach to probing nitrogen-fixing bacteria in soil communities by resonance raman spectroscopy with $^{15}N_2$ labeling. Anal. Chem. *90*, 5082–5089. https://doi.org/10.1021/acs.analchem.7b05080

Dieterich, D.C., Link, A.J., Graumann, J., Tirrell, D.A., and Schuman, E.M. (2006). Selective identification of newly synthesized proteins in mammalian cells using bioorthogonal noncanonical amino acid tagging (BONCAT). Proc. Natl. Acad. Sci. U.S.A. *103*, 9482–9487.

Fierer, N. (2017). Embracing the unknown: disentangling the complexities of the soil microbiome. Nat. Rev. Microbiol. *15*, 579–590. https://doi.org/10.1038/nrmicro.2017.87

Gilbert, J.A., and Henry, C. (2015). Predicting ecosystem emergent properties at multiple scales. Environ. Microbiol. Rep. *7*, 20–22. https://doi.org/10.1111/1758-2229.12258

Gonzalez, A., King, A., Robeson, M.S., Song, S., Shade, A., Metcalf, J.L., and Knight, R. (2012). Characterizing microbial communities through space and time. Curr. Opin. Biotechnol. *23*, 431–436. https://doi.org/10.1016/j.copbio.2011.11.017

Gupta, V.V., and Germida, J.J. (2015). Soil aggregation: Influence on microbial biomass and implications for biological processes. Soil Biol. Biochem. *80*, A3–A9.

Hatzenpichler, R., Scheller, S., Tavormina, P.L., Babin, B.M., Tirrell, D.A., and Orphan, V.J. (2014). In situ visualization of newly synthesized proteins in environmental microbes using amino acid tagging and click chemistry. Environ. Microbiol. *16*, 2568–2590. https://doi.org/10.1111/1462-2920.12436

Hekstra, D.R., and Leibler, S. (2012). Contingency and statistical laws in replicate microbial closed ecosystems. Cell *149*, 1164–1173. https://doi.org/10.1016/j.cell.2012.03.040

Hendershot, J.N., Read, Q.D., Henning, J.A., Sanders, N.J., and Classen, A.T. (2017). Consistently inconsistent drivers of microbial diversity and abundance at macroecological scales. Ecology *98*, 1757–1763. https://doi.org/10.1002/ecy.1829

Huang, W.E., Griffiths, R.I., Thompson, I.P., Bailey, M J., and Whiteley, A.S. (2004). Raman microscopic analysis of single microbial cells. Anal. Chem. *76*, 4452–445.

Huang, W.E., Stoecker, K., Griffiths, R., Newbold, L., Daims, H., Whiteley, A.S., and Wagner, M. (2007). Raman-FISH: combining stable-isotope Raman spectroscopy and fluorescence in situ hybridization for the single cell analysis of identity and function. Environ. Microbiol. *9*, 1878–1889.

Hurlbert, S.H. (1984). Pseudoreplication and the design of ecological field experiments. Ecol. Monogr. *54*, 187-211.

Huys, G.R., and Raes, J. (2018). Go with the flow or solitary confinement: a look inside the single-cell toolbox for isolation of rare and uncultured microbes. Curr. Opin. Microbiol. *44*, 1–8.

Jansson, J.K., and Hofmockel, K.S. (2018). The soil microbiome-from metagenomics to metaphenomics. Curr. Opin. Microbiol. *43*, 162–168.

Legendre, P., and Gauthier, O. (2014). Statistical methods for temporal and space-time analysis of community composition data. Proc. Biol. Sci. *281*, 20132728. https://doi.org/10.1098/rspb.2013.2728

Lipson, D.A., and Schmidt, S.K. (2004). Seasonal changes in an alpine soil bacterial community in the colorado rocky mountains. Appl. Environ. Microbiol. *70*, 2867–2879.

Levin, S.A. (1992). The problem of pattern and scale in ecology: the Robert H. MacArthur award lecture. Ecology *73*, 1943–1967.

Marco, D. (2017). Integration of Ecology and Environmental Metagenomics Conceptual and Methodological Frameworks. Curr. Issues Mol. Biol. *24*, 1–16. https://doi.org/10.21775/cimb.024.001

Medema, M.H. (2018). computational genomics of specialized metabolism: from natural product discovery to microbiome ecology. mSystems *3*, e00182–17. https://doi.org/10.1128/mSystems.00182-17

Meiring, T.L., Bauer, R., Scheepers, I., Ohloff, C., Tuffin, I.M., and Cowan, D.A. (2011). Metagenomics and beyond: current approaches and integration with complementary technologies. In Metagenomics: Current Innovations and Future Trends, Marco, D., ed. (Caister Academic Press, Norfolk), pp. 1–19.

Mikola, J., Bardgett, R.D., and Hedlund, K. (2002). Biodiversity, ecosystem functioning and soil decomposer food webs. In Biodiversity and Ecosystem Functioning: Synthesis and Perspectives, Loreau, M., Naeem, S., and Inchausti, P., eds. (Oxford University Press, Oxford), pp. 169–180.

Nadell, C.D., Drescher, K., and Foster, K.R. (2016). Spatial structure, cooperation and competition in biofilms. Nat. Rev. Microbiol. *14*, 589–600. https://doi.org/10.1038/nrmicro.2016.84

O'Brien, S.L., Gibbons, S.M., Owens, S.M., Hampton-Marcell, J., Johnston, E.R., Jastrow, J.D., Gilbert, J.A., Meyer, F., and Antonopoulos, D.A. (2016). Spatial scale drives patterns in soil bacterial diversity. Environ. Microbiol. *18*, 2039–2051. https://doi.org/10.1111/1462-2920.13231

O'Neill, R.V., DeAngelis, D.L., Waide, J.B., and Allen, T.F.H. (1986). A hierarchical concept of ecosystems (Princeton University Press, Princeton, New Jersey).

Paul, D., Kumar, S., Mishra, M., Parab, S., Banskar, S., and Shouche, Y.S. (2018). Molecular genomic techniques for identification of soil microbial community structure and dynamics. In Advances in Soil Microbiology: Recent Trends and Future Prospects, Volume 1, Adhya T.K., Lal B., Mophapatra B., Paul D., and Das, S., eds (Springer, Singapore), pp. 9–33.

Prosser, J.I. (2010). Replicate or lie. Environ. Microbiol. *12*, 1806–1810.

Prosser, J.I. (2015). Dispersing misconceptions and identifying opportunities for the use of 'omics' in soil microbial ecology. Nat. Rev. Microbiol. *13*, 439–446. https://doi.org/10.1038/nrmicro3468

Prosser, J.I. (2017). Advances in the study of the microbiome. Microbiology Today *44*, 95–96.

Razanamalala, K., Razafimbelo, T., Maron, P.A., Ranjard, L., Chemidlin, N., Lelièvre, M., Dequiedt, S., Ramaroson, V.H., Marsden, C., Becquer, T., *et al.* (2018). Soil microbial diversity drives the priming effect along climate gradients: a case study in Madagascar. ISME J. *12*, 451–462. https://doi.org/10.1038/ismej.2017.178

Reim, A., Lüke, C., Krause, S., Pratscher, J., and Frenzel, P. (2012). One millimetre makes the difference: high-resolution analysis of methane-oxidizing bacteria and their specific activity at the oxic-anoxic interface in a flooded paddy soil. ISME J. *6*, 2128–2139. https://doi.org/10.1038/ismej.2012.57

Rillig, M.C., Muller, L.A., and Lehmann, A. (2017). Soil aggregates as massively concurrent evolutionary incubators. ISME J. *11*, 1943–1948. https://doi.org/10.1038/ismej.2017.56

Rosenzweig, M.L., and Ziv, Y. (1999). The echo pattern of species diversity: pattern and processes. Ecography *22*, 614–628.

Schadt, C.W., Martin, A.P., Lipson, D.A., and Schmidt, S.K. (2003). Seasonal dynamics of previously unknown fungal lineages in tundra soils.Science *301*, 1359–1361. https://doi.org/10.1126/science.1086940

Schneider, D.C. (1994). Quantitative Ecology: Spatial and Temporal Scaling (Academic Press, San Diego, CA, USA).

Shi, Y., Grogan, P., Sun, H., Xiong, J., Yang, Y., Zhou, J., and Chu, H. (2015). Multi-scale variability analysis reveals the importance of spatial distance in shaping Arctic soil microbial functional communities. Soil Biol. Biochem. *86*, 126–134.

Singer, E., Wagner, M., and Woyke, T. (2017). Capturing the genetic makeup of the active microbiome in situ. ISME J. *11*, 1949–1963. https://doi.org/10.1038/ismej.2017.59

Stempfhuber, B.H.J. (2016). Drivers for the performance of nitrifying organisms and their temporal and spatial interaction in grassland and forest ecosystems (Doctoral dissertation, Technische Universität München, München).

Štursová, M., Bárta, J., Šantrůčková, H., and Baldrian, P. (2016). Small-scale spatial heterogeneity of ecosystem properties, microbial community composition and microbial activities in a temperate mountain forests oil. FEMS Microbiol. Ecol. 92, fiw185.

Thompson, L.R., Sanders, J.G., McDonald, D., Amir, A., Ladau, J., Locey, K.J., Prill, R.J., Tripathi, A., Gibbons, S.M., Ackermann, G., *et al.* (2017). A communal catalogue reveals Earth's multiscale microbial diversity. Nature *551*, 457–463.

Vos, M., Wolf, A.B., Jennings, S.J., and Kowalchuk, G.A. (2013). Micro-scale determinants of bacterial diversity in soil. FEMS Microbiol. Rev. 37, 936–954. https://doi.org/10.1111/1574-6976.12023

Wawrik, B. (2014). Stable isotope probing the N cycle: current applications and future directions. In Metagenomics of the Microbial Nitrogen Cycle, Theory, Methods and Applications, Marco, D., ed. (Caister Academic Press, Norfolk), pp. 87–110.

Wiens, J.A. (1989). Spatial scaling in ecology. Funct. Ecol. *3*, 385–397.

Willerslev, E., Hansen, A.J., Rønn, R., Brand, T.B., Barnes, I., Wiuf, C., Gilichinsky, D., Mitchell, D., and Cooper, A. (2004). Long-term persistence of bacterial DNA. Curr. Biol. *14*, R9–10.

Zhou, J., He, Z., Yang, Y., Deng, Y., Tringe, S.G., and Alvarez-Cohen, L. (2015). High-throughput metagenomic technologies for complex microbial community analysis: open and closed formats. MBio *6*, e02288–14. https://doi.org/10.1128/mBio.02288-14

Contamination Issues in Microbiome Sequencing Studies

2

Sharon Bewick[1,2*], David Karig[3,4] and William F. Fagan[1]

[1]Department of Biology, University of Maryland, College Park, MD, USA.
[2]Department of Biological Sciences, Clemson University, Clemson, SC, USA.
[3]Research and Exploratory Development Department, Johns Hopkins University Applied Physics Laboratory, Laurel, MD, USA.
[4]Department of Bioengineering, Clemson University, Clemson, SC, USA.

*Correspondence: sharon_bewick@hotmail.com

https://doi.org/10.21775/9781912530021.02

Abstract

Contamination in microbiome sequencing studies is a well-known but difficult to address problem. Unlike culture experiments, in which only living cell contamination is of consequence, sequencing analyses can be disrupted by dead cells or even small pieces of non-viable DNA. Although all sequencing studies face contamination issues, these issues are likely to be more or less important, depending on the microbiome under examination. Systems characterized by low input samples, for example the human skin microbiome, may be highly susceptible to contamination. By contrast, systems characterized by high input samples, for example the human gut microbiome, may be less influenced. Here, we use a simulation model to examine when contamination is likely to be problematic, relating the degree to which sequencing results are influenced by contamination to sample characteristics such as DNA input, as well as ecological characteristics such as taxon diversity. We also study how well methods such as thresholding work to differentiate target taxa from contaminants. This work illustrates some of the challenges associated with the collection and interpretation of sequencing studies when used to understand the composition and behaviour of complex microbial communities.

Introduction

The past decade has seen a rapid growth in the use of sequencing technology for characterization of microbial communities. A key benefit of the sequencing approach is that it can detect unculturable bacteria, providing a better picture of microbiome diversity across a wide array of habitats. Unfortunately, this benefit comes with a cost – increased detection extends to contaminants, including living and dead microbes and free DNA that enter the sample from a range of sources, for example polymerase chain reaction (PCR) tubes (Schmidt *et al.*,

1995), DNA extraction kits (Evans *et al.*, 2003) and even bacteria from researchers involved in sample preparation (Salter *et al.*, 2014; Weiss *et al.*, 2014). Indeed, it has been suggested that non-endogenous microbial DNA contaminates every microbiome dataset to at least some extent (Weiss *et al.*, 2014). Unfortunately, we have very limited knowledge of the spectrum of potential taxa that are likely to contaminate sequencing studies (Lusk, 2014). This makes it almost impossible to know which taxa originated from the target environment and which were accidentally introduced during a sample preparation step.

Although contamination has always been a concern in microbiome sequencing studies, issues with contaminants are being brought to the forefront by the extension of sequencing techniques to new and more challenging microbiomes. Early microbiome work focused on environments with high biomass, for example the human gut (Gill *et al.*, 2006). In these systems, contamination is likely to be of minimal consequence, because the high biomass of the sample itself almost certainly overwhelms contaminating background microbial DNA (Weiss *et al.*, 2014). More recently, however, there has been growing interest in studying systems with low biomass, for example the human skin microbiome (Grice *et al.*, 2009) or lung microbiome (Erb-Downwards *et al.*, 2011). In these systems, the biomass from the target environment is smaller, meaning that there is greater potential for contaminants to contribute a larger fraction of the sequences (Weiss *et al.*, 2014).

Unfortunately, contaminants in sequencing studies are not just a nuisance. They can fundamentally impact biological conclusions and, in some cases, even suggest incorrect health implications. Salter *et al.* (2014), for example, showed how contaminant operational taxonomic units (OTUs) from a DNA extraction kit could result in the misleading conclusion that the nasopharyngeal microbiome changed with age. Likewise, Lauder *et al.* (2016) demonstrated how bacteria attributed to the placental microbiome – a microbiome implicated in reproductive health and disease – could not be reasonably distinguished from contamination introduced during DNA purification. In both cases, analysis was complicated because the environments of interest provided low biomass samples that were highly susceptible to even small levels of contamination.

A range of techniques has been suggested for minimizing contaminants in sequencing studies. Exposing PCR reagents to UV radiation, for example, can inactivate free DNA (Fox *et al.*, 1991). Similarly, positive pressure laboratory ventilation systems can help to prevent diffusion of microbes and other DNA into room air (Lusk, 2014). Some researchers advocate even stricter protocols, for instance the use of clean-suits, gloves, facemasks and bleach, and UV radiation for cleaning equipment (Weiss *et al.*, 2014). Although these precautions can certainly reduce contamination levels, they do not appear to be sufficient to eliminate contaminants completely (Lusk, 2014; Champlot *et al.*, 2010; Gill *et al.*, 2000; Corless *et al.*, 2000). An alternative approach is to compile lists of common contaminants (Salter *et al.*, 2014; Laurence *et al.*, 2014) and then to ignore these taxa or, at the very least, treat them with a high degree of scepticism, in all future sequencing studies. Sometimes, however, common sequencing contaminants may also be integral members of the microbiome being studied. Thus, it is unwise to indiscriminately ignore all OTUs that are either common contaminants or else are found in concurrent controls. This is particularly true for systems like human oral and skin microbiomes. Indeed, members of these communities are likely to show up in controls solely because samples are routinely exposed to the oral and skin microbiomes on technicians during sample preparation (Weiss *et al.*, 2014).

Except in rare cases, the total input from contaminant DNA will be small compared with the total input from the target environment, even for samples with characteristically low biomass. As a consequence, it would be highly unlikely for the most abundant OTUs in a sample to be derived solely from contamination. Indeed, problems with contamination are more likely to appear in resolving and interpreting the tail of rare organisms that characterize most complex microbiomes. Consequently, beyond trying to prevent contamination or compile lists of common contaminants, an alternative strategy for dealing with contamination is to attempt to delineate exactly how and where in sequencing data contamination is likely to be problematic. In this chapter, we use assumptions about the distributions of contaminant cells and microbiome constituents in order to explore the degree to which microbiome datasets are reliable as a function of the relative degree of contamination, and the OTU diversity of the target microbiome and contaminant sources. Specifically, we consider how far into the rare tail OTUs can be reliably attributed to the target environment, and we use this to establish thresholds for data interpretation.

Method

Sample

We begin by assuming that there are two bacterial communities entering any sample – the community from the target environment and the community from the contaminant source. Therefore, the OTU diversity in the sample, S_{sample}, is a combination of the OTU diversity in the target environment, S_{target}, and the OTU diversity in the contamination, $S_{\text{contaminant}}$. There may, however, be some OTU overlap between these two sources, thus:

$$S_{\text{target}} = (1-\theta)S_{\text{sample}} + \varphi S_{\text{sample}} \quad (2.1a)$$

$$S_{\text{contaminant}} = \theta S_{\text{sample}} \quad (2.1b)$$

where θ is the fraction of sample OTU diversity that is present in the contaminant source and φ is the fraction of sample OTU diversity that is present in *both* the contaminant source and the target environment. For $\varphi > 0$, we assume that there is no correlation between OTU abundance in the target environment and the contamination (i.e. OTUs that are abundant in the target environment are no more or less likely to be abundant in the contamination than are OTUs that are rare in the target environment).

In reality, there may be several contaminant communities, for example contaminants from the room air, contaminants from the water source, and even contaminants from the sequencing kits. The assumption of a single contaminant community is, thus, a first approximation. We assume that both the target community and the contaminant community are characterized by independent log-series OTU abundance distributions. We select log-series distributions because this is a commonly proposed distribution for ecological communities (Boswell and Patil, 1971; Hill and Hamer, 1998; Kempton and Taylor, 1974), including those that arise from sampling processes (e.g. dispersal to an island, sample collection) (Williamson and Gaston, 2005; Fisher *et al.*, 1943; Kendall, 1948; Bewick *et al.*, 2015; May, 1975).

To generate a community with a log-series abundance distribution, we assume that the target (alternatively, contaminant) component of the sample contains $w_{\text{target(contaminant)}}$ nanograms of bacterial DNA. For bacteria with an average genome size of μ base pairs (bps), this corresponds to $C_{\text{target(contaminant)}} = w_{\text{target(contaminant)}} \cdot 9 \times 10^{11}/\mu$ bacterial cells. Given the number of cells, C, and the OTU diversity, S, the log-series distribution can be fully characterized by the parameter x, where x is determined as follows:

$$x = \frac{C}{C+\alpha} \text{ and } \alpha = \frac{S}{-\log(1-x)} \tag{2.2}$$

Equation 2.2 can be solved numerically for x. Using x, we generate a series of log-series distributed random numbers according to the LS algorithm in Kemp (1981) and Bewick *et al.* (2017). Because of the way this algorithm is implemented, the total number of OTUs and cells in the series vary stochastically and can deviate slightly from their input values. Consequently, we repeat the LS algorithm until the number of OTUs in the series is exactly equal to S and until the number of cells in the series falls within plus or minus the mean OTU population size, v, where:

$$v = -\frac{1}{\log(1-x)}\frac{x}{1-x} \tag{2.3}$$

At this point, the simulation is stopped, with the abundance of each OTU given by the series of randomly generated population sizes.

16S copy number

For each of the OTUs in the sample, we assign a random 16S copy number, λ. Copy numbers are selected by generating random numbers from the empirical probability distribution in Větrovský and Baldrian (2013).

Sequencing

To mimic the sequencing process, we assume p rounds of PCR, with an efficiency, ε. For the sake of simplicity and computational time, we do not consider any stochasticity in the sequencing step; hence, after amplification, the number of reads, R, representing an OTU with abundance n is:

$$R = (1+\varepsilon)^p \lambda n \tag{2.4}$$

Sequencing is then accomplished by randomly drawing (without replacement) and assigning taxonomy to D reads from the pool of reads for all OTUs, where D is the sequencing depth. We assume a 1 : 1 ratio of distinct 16S sequences to OTUs, and perfect OTU assignment. This is an oversimplification, because 16S sequences are typically clustered based on sequence similarity with different clusters assigned to different OTUs.

Parameters

Unless stated otherwise, we assume that $p = 38$, $\mu = 4 \times 10^6$ bps and $\varepsilon = 0.50$ which, for a sample with 1 ng of bacterial DNA, yields an approximately 400-fold increase in nanograms of DNA during the amplification step (see Appendix 1). This is consistent with common 16S sequencing protocols. For the sake of simplicity, we assume that there is a full 1 ng of bacterial DNA from the target environment and that there are, in addition, varying levels of contaminant DNA ranging from $w_{contaminant} = 0.001$ ng to $w_{contaminant} = 0.1$ ng. We consider three different diversity levels, $S_{sample} = 100$, 200 and 300, which represent low, intermediate and high diversity systems (we avoid $S_{sample} > 300$ primarily because of computational time, realizing that there are microbiomes with greater diversity than we have considered). We assume different partitioning of diversity among target and contaminant sources ranging from $\theta = 0.05$ to $\theta = 0.95$ and from $\varphi = 0$ to $\varphi = 0.25$. Finally, to explore the role of sequencing depth, we consider three protocols with $D = 1 \times 10^4$, 1×10^5 and 1×10^6 reads.

Results

To understand the impact of contaminants on microbiome community characterization, we consider the probability that an OTU is not a member of the target community as a function of OTU rank abundance. Ideally, all taxa from the target environment should have lower rank abundances (higher absolute abundances) than all taxa solely found in the contaminant source, making it easy to separate target organisms from contaminants. This, however, will not always be the case. In general, there are two reasons why a contaminant might have a lower rank abundance than a target organism. First, contamination may be sufficiently high, such that OTUs that are abundant in the contaminant source contribute more cells to the sample than OTUs that are rare in the target environment. Second, even when a contaminant OTU contributes fewer cells to the sample, it may still appear at a lower rank abundance because of stochasticity in the sampling process (i.e. sequencing stochasticity).

Sequencing depth

When sequencing stochasticity is the cause of poor separation between OTUs from the contaminant source and the target environment, deeper sequencing can improve results. Fig. 2.1 shows the probability of an OTU being solely from the contaminant source as a function of its rank abundance percentile (i.e. rank abundance, rescaled between 0 and 100) in the sequencing data. For the sake of comparison, we show curves using rank abundance percentiles based on cell counts as well. These latter curves remove errors/stochasticity as a result of sequencing and thus represent the 'true' probabilities that a population with a particular rank abundance is from contamination versus the target environment. All else being equal, deeper sequencing results in greater confidence that OTUs at higher rank abundances (i.e. more rare) are from the target environment. This is particularly true for samples with higher diversity, where deeper sequencing is necessary to successfully resolve OTU relative abundances (compare, for example, the light grey and black curves in Fig. 2.1B).

At very low sequencing depths ($D = 1 \times 10^4$ reads; see Fig. 2.1A), sequence-based OTU curves (solid lines) for low (light grey), intermediate (dark grey) and high (black) diversity samples coincide and strongly deviate from corresponding cell-based curves (dashed lines). This suggests that sequencing error dominates and is the limiting factor for resolving contaminants from target organisms. For the example shown in Fig. 2.1A, for instance, sequencing depth restricts data reliability to the top ≈20% of the most abundant OTUs. That is, OTUs with lower abundance may or may not be from the target environment. At intermediate sequencing depths ($D = 1 \times 10^5$ reads; see Fig. 2.1B), the sequence-based and cell-based curves are nearly converged for the low diversity sample (light grey). However, these curves remain distinct for the intermediate (dark grey) and high (black) diversity samples. Thus, for

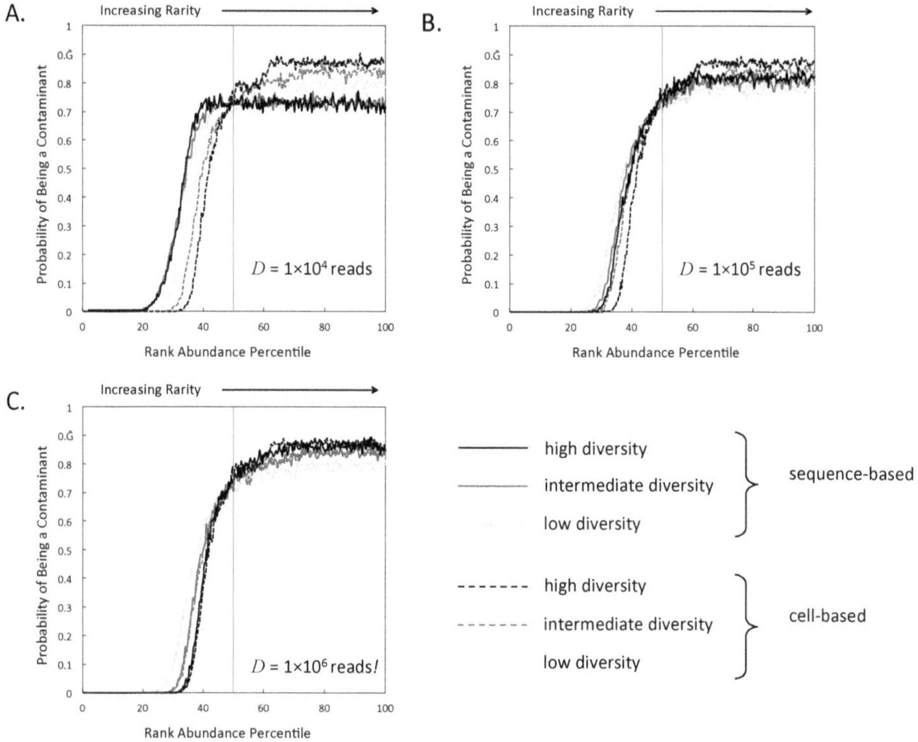

Figure 2.1 The probability that an OTU comes from contamination as a function of the OTU rank abundance percentile (i.e. rank abundance rescaled between 0 and 100) for sequencing depths of (A) $D = 1 \times 10^4$, (B) $D = 1 \times 10^5$ and (C) $D = 1 \times 10^6$. Solid lines are based on sequenced reads, whereas dashed lines are based on actual cell counts in the sample. Black, dark grey and light grey lines are used for samples with high (300 OTUs), intermediate (200 OTUs) and low (100 OTUs) diversity respectively. The red vertical line represents the ideal transition point between target and contaminant OTUs. For all figures, we assume that there is 1 ng of bacterial DNA from the target environment, and 0.001 ng of bacterial DNA from the contaminant source, and that the target and contaminant communities are equally diverse, with no OTU overlap ($\theta = 0.5$, $\phi = 0$). Curves are averages over 1000 simulations.

higher diversity communities, sequencing error remains a significant obstacle to resolving contaminants from target organisms even at $D = 1 \times 10^5$. At much higher sequencing depths ($D = 1 \times 10^6$ reads; see Fig. 2.1C), the sequence-based and cell-based curves have converged, even for the highest diversity sample. At this point, further sequencing will not be helpful, because sampling error is no longer the greatest obstacle to contaminant resolution. Rather, resolution issues are a result of there being too much contamination in the sample. In this case, the only option is to reduce contaminant input. Interestingly, for the same level of contamination ($w_{\text{contaminant}} = 0.001$ ng), and the same partitioning of OTU diversity between the target environment and contamination ($\theta = 0.5$, $\varphi = 0$), a larger percentage and, by extension, a larger absolute number of OTUs can be trusted in the more diverse samples. For the remainder of our results, we focus on $D = 1 \times 10^6$ reads, which effectively removes sequencing error for the sample diversities that we consider.

Contaminant input

When sequencing error has been ruled out as the source of poor resolution between contaminant and target DNA, improved microbiome characterization requires increased effort to reduce contamination. For the analysis in the previous section, we assumed very low levels of contaminant input ($w_{contaminant} = 0.001$ ng), such that there was 1000-fold less DNA from the contaminant source than from the target environment. In reality, however, contamination can be much higher. One way to estimate the degree of contamination is to compare DNA quantification in samples and controls. Fig. 2.2A shows DNA yield for typical skin swabs from human foreheads, along with DNA yield for controls collected at the same time as the samples. In this experiment, DNA from the target environment was only ≈10-fold higher than DNA from contaminants, even with the somewhat overly optimistic assumption that non-bacterial DNA comprises similar fractions in both samples and controls. Skin samples, of course, represent a particular challenge in terms of low input. Nevertheless, this gives a sense of the level of contamination likely in typical microbiome work.

Fig. 2.2B shows the effect of contamination on our ability to characterize the target microbiome community. Specifically, we plot the rank abundance percentile of the rarest OTU with < 1% probability of being a contaminant as a function of contaminant input. What is clear is that there is an exponential decrease in the percentage of OTUs that can be reliably attributed to the target environment with increasing levels of contamination. Indeed, whereas ≈33% of OTUs can be reasonably attributed to the target environment when contaminant input is 1000-fold less than target input, only ≈3% of OTUs are trustable when contaminant input is only 10-fold less than target input. This emphasizes the importance of reducing contamination as much as possible during sample collection and preparation. Although higher diversity communities enable trustworthy identification of OTUs at higher rank abundance percentiles at all contaminant levels, this benefit becomes less pronounced with increasing contamination.

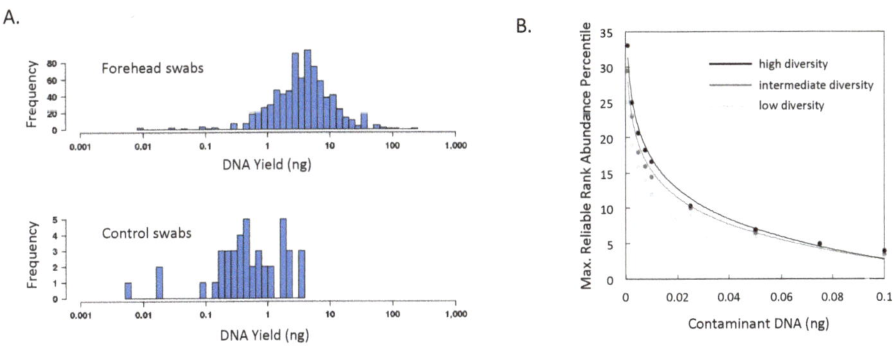

Figure 2.2 (A) Histograms of DNA yield from forehead skin microbiome samples and controls. (B) Rank abundance percentile of the rarest OTU with <1% probability of being a contaminant as a function of nanograms of contaminant DNA for samples with high (300 OTUs, black line), intermediate (200 OTUs, dark grey line) and low (100 OTUs, light grey line) diversity. For all three curves, we assume that there is 1 ng of bacterial DNA from the target environment, that the sequencing depth is $D = 1 \times 10^6$ reads and that the target and contaminant communities are equally diverse, with no OTU overlap ($\theta = 0.5$, $\phi = 0$). Curves are averages over 1000 simulations and logarithmic trend lines are provided as guides.

OTU diversity

In the above analysis, we have assumed that OTU diversity is equally partitioned between the target environment and the contamination. However, depending on the microbiome being studied and the contaminant source, the target environment may contribute a smaller or larger fraction of OTUs to the sample. Fig. 2.3A shows a histogram of the average ratio of total sample OTU diversity to total control OTU diversity for 51 forehead microbiome samples paired with controls. Even for this relatively homogeneous set of samples, there is wide variation in the OTU ratios between samples and controls (minimum diversity ratio 0.40 vs. maximum diversity ratio 6.82). Fig. 2.3B shows the effect of different partitioning of OTU diversity on our ability to characterize a microbial community. Not surprisingly, when 100% of OTU diversity is found in the contaminant source (and assuming no OTU overlap between the target environment and contamination), there are no OTUs that can be reliably attributed to the target environment.

Indeed, in this limit, there are no OTUs in the target environment at all. However, as more and more OTUs are found in the target environment (and by extension, fewer and fewer OTUs are found in the contaminant source), it becomes possible to ascribe an increasing number of rare OTUs to the target environment. Interestingly, however, the relationship is non-monotonic (i.e. is hump-shaped). Thus, when the contaminant source becomes very OTU poor, the maximum reliable rank abundance decreases again. Although somewhat counterintuitive, the explanation is straightforward. When the contaminant source has very few OTUs, one or a few populations account for the majority of DNA input. As a result, even for low contaminant input, these populations can achieve abundances that rival some of the rarer OTUs from an OTU rich target environment. In general, the curves in Fig. 2.3B

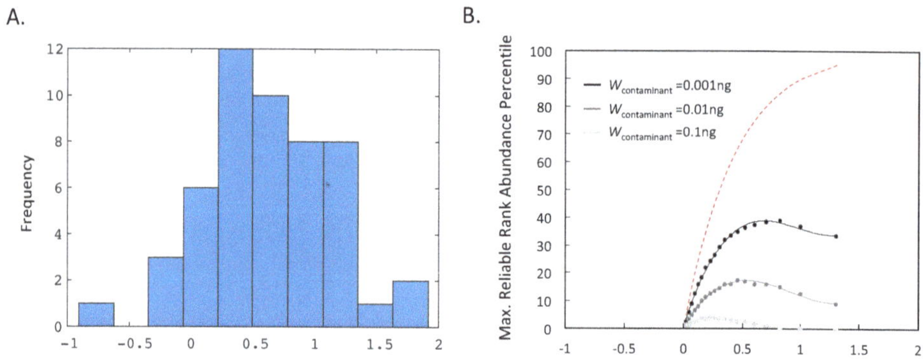

Figure 2.3 (A) Histograms of the ratios of OTU diversity in samples versus controls for forehead skin swabs. (B) Rank abundance percentile of the rarest OTU with <1% probability of being a contaminant as a function of the ratio of OTU diversity in the sample versus controls assuming low contamination ($w_{contaminant}$=0.001, black line), intermediate contamination ($w_{contaminant}$=0.01, dark grey line) and high contamination ($w_{contaminant}$=0.1, light grey line). For all three curves, we assume that there is 1 ng of bacterial DNA from the target environment, that the sequencing depth is $D = 1 \times 10^6$ reads, and that there are a total of 200 OTUs in the sample (environmental + contaminant OTUs), with no OTU overlap ($\phi = 0$). The dashed red line is the theoretical limit to the maximum reliable rank abundance percentile (i.e. assuming perfect separation between target environment and contamination). Curves are averages over 1000 simulations and polynomial trend lines are provided as guides.

peak at lower contaminant diversity for higher contaminant input. In other words, when there is a lot of contamination, it is often better if this contamination comes from a wide range of OTUs because, in this case, no single population will be overly abundant. However, when there is very little contamination, even if this contamination is dominated by one or a few OTUs, these OTUs are less likely to achieve abundances rivalling the more abundant populations from the target environment.

OTU overlap

Thus far, we have assumed that there is no overlap between the OTUs that occur in the target environment and the OTUs that occur in the contamination (i.e. $\varphi = 0$). If this were always the case, however, the issue of resolving contaminant DNA from target DNA would be trivial – any OTUs that occur in controls could be ignored, leaving only OTUs that are solely present in the sample. Unfortunately, for most samples, some degree of overlap between OTUs from the contaminant source and the target environment is likely (i.e. $\varphi > 0$). This could occur for several reasons. First, contamination may stem from a microbiome similar to the source itself. This is particularly true of certain human microbiomes where, for instance, room air can be contaminated with skin and/or oral bacteria from the people working on sample preparation. It is also true for certain freshwater microbiomes where water used for sample preparation could harbour the same taxa as the target environment being investigated. A second, and more insidious possibility is that microbes in the contaminant source differ from microbes in the target environment, but that the taxonomic resolution achieved through sequencing is not sufficient to differentiate these taxa. *Pseudomonas* spp., for example, are common contaminants that arise during sample preparation. However, *Pseudomonas* can also be cultured directly from human skin samples, suggesting that this genus is also present in many human microbiomes. While the precise taxonomy of the *Pseudomonas* spp. from contaminant sources and the human skin microbiome likely differ, 16S sequencing using short variable regions will typically be insufficient to distinguish the different species/strains.

Fig. 2.4 shows the impact of increasing OTU overlap between the target environment and the contaminant source (i.e. increasing φ). Notably, complete OTU separation represents a worst-case scenario, with increasing overlap resulting in increasing confidence in rarer OTUs. Sadly, however, this improvement is most apparent when contaminant input is low, and becomes almost negligible at high contaminant levels. Nevertheless, it suggests that estimates for the maximum reliable rank abundances made by assuming zero OTU overlap give conservative, lower bounds on the percentage of organisms that can be trusted from sequencing data.

Relative abundance thresholds

Throughout the previous sections, we have framed all results in terms of rank abundance percentiles. This gives a sense of the fraction of OTUs that can be trusted in any particular experiment. Often, however, the decision to trust a particular OTU in a particular dataset is based on a relative abundance threshold. To explore this approach for interpreting metagenomics data, in Fig. 2.5 we recreate Figs. 2.2B and 2.3B in terms of OTU relative abundance, rather than rank abundance.

As with rank abundances, there is no single relative abundance threshold that provides a reliable cut-off between organisms from the target environment and contaminants. Indeed,

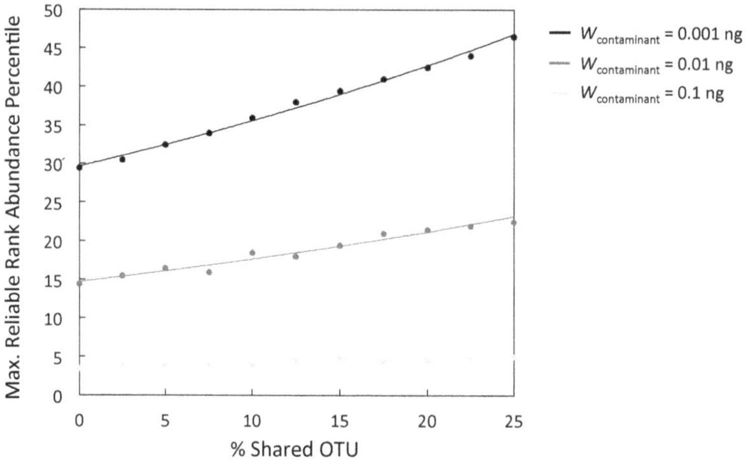

Figure 2.4 Rank abundance percentile of the rarest OTU with <1% probability of being a contaminant as a function of the percentage of OTUs shared between the target environment and the contaminant source assuming low contamination ($w_{contaminant}$=0.001, black line), intermediate contamination ($w_{contaminant}$=0.01, dark grey line) and high contamination ($w_{contaminant}$=0.1, light grey line). For all three curves, we assume that there is 1 ng of bacterial DNA from the target environment, that the sequencing depth is $D=1\times10^6$ reads and that there are a total of 200 OTUs in the sample (environmental + contaminant OTUs). Curves are averages over 1000 simulations and exponential trend lines are provided to guide the eye.

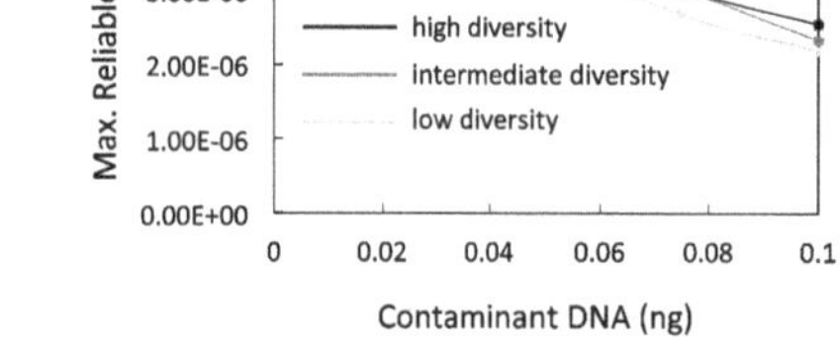

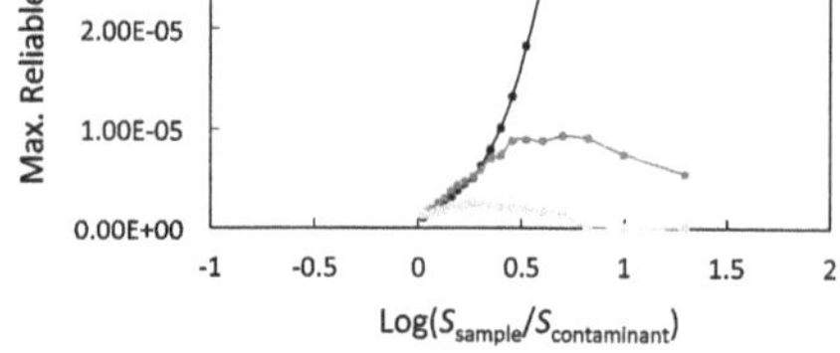

Figure 2.5 (A) Relative abundance of the rarest OTU with <1% probability of being a contaminant as a function of nanograms of contaminant DNA for samples with high (300 OTUs, black line), intermediate (200 OTUs, dark grey line) and low (100 OTUs, light grey line) diversity and assuming a low contaminant level of $w_{contaminant}$=0.001. (B) Relative abundance of the rarest OTU with <1% probability of being a contaminant as a function of the ratio of OTU diversity in the sample versus controls assuming low contamination ($w_{contaminant}$=0.001, black line), intermediate contamination ($w_{contaminant}$=0.01, dark grey line) and high contamination ($w_{contaminant}$=0.1, light grey line) and assuming an intermediate sample diversity of 200 OTUs. For all curves, we assume that there is 1 ng of bacterial DNA from the target environment, that the sequencing depth is $D=1\times10^6$ reads, and that the target and contaminant communities are equally diverse, with no OTU overlap (θ=0.5, ϕ=0). Curves are averages over 1000 simulations.

even for the same contamination level, this cut-off can vary over an order of magnitude based on the OTU diversities of the target environment and contaminant source (see Fig. 2.5B). Interestingly, in contrast to the curves in Fig. 2.2B, which are monotonically decreasing, the curves in Fig. 2.5A are non-monotonic, increasing with contamination at low contaminant levels, but then decreasing again at high contaminant levels. This is because, at low contamination, the threshold for reliability lies in the rare tail of the species abundance distribution (SAD). As a result, small increases in contamination cause a dramatic reduction in the number of OTUs that can be trusted (see Fig. 2.2B), with only those OTUs that have high enough abundance to significantly outnumber abundant contaminants being trustworthy. By contrast, at high contamination, only the very abundant target organisms can be trusted anyways (see Fig. 2.2B). As a result, the primary effect of increasing contamination at this point is to reduce the total relative abundance of the highly abundant target OTUs.

Discussion

In this chapter, we have examined the impact of contamination on data from amplicon sequencing studies. Specifically, we have capitalized on the fact that many biological communities (Boswell and Patil, 1971; Hill and Hamer, 1998; Kempton and Taylor, 1974), and particularly those that arise from a sampling process (Williamson and Gaston, 2005; Fisher *et al.*, 1943; Kendall, 1948; May, 1975), exhibit log-series abundance distributions. Because the log-series distribution is fully defined by two parameters, the total number of cells in a community, along with total OTU diversity, fully specifies the shape of the distribution. By applying this fact to microbes from both a contaminant pool and a target environment, we are able to compare population abundances between these two input sources. Notably, contaminant diversity can be estimated based on OTU counts in controls, whereas total cell counts can be estimated (at least loosely) based on DNA quantification of controls and samples, along with average bacterial genome sizes. That said, for samples where a large amount of DNA is expected from non-microbial sources, paired shotgun sequencing may be necessary to get reasonable estimates of the fraction of DNA that can be attributed to microbial populations.

As expected, and suggested in previous discussions of contamination (Weiss *et al.*, 2014), higher ratios of contaminant-to-target DNA rapidly degrade the information available from sequencing data (notice that, although we always use 1 ng of input from the target environment, what matters is the ratio of contamination to target DNA; thus, low biomass samples can be thought of as samples with higher contamination). Indeed, there is an exponential decrease in the number of reliable OTUs as a function of contaminant input (see Fig. 2.2B). This is one key reason why low biomass samples present such a challenge. In our study, we were able to quantify the approximate relationship between trustable OTUs and contamination for a given level of sample diversity. Shockingly, to trust even approximately one-third of the OTUs in a dataset with 100–300 OTUs, contamination must be ≈1000-fold lower than input from the target environment. For traditionally low input samples, for example the human skin microbiome, this is a lofty goal (see Fig. 2.2A). Although this particular estimate assumes equal partitioning of OTU diversity between the target environment and the contaminant source and no OTU overlap among the two, gains as a result of increasing the diversity in the target environment relative to contamination are modest, as are gains resulting from OTU overlap (see Figs. 2.3B and 2.4). For the significantly higher contamination levels typical of low input samples, the outlook is even worse. Indeed, when contaminant DNA is only 10-fold lower than input from the target environment, only ≈3% of OTUs are trustable, which is typically fewer than 10 genera.

Just as reliable identification of target OTUs is challenging, so too is reliable identification of contaminants. In some cases, deeper sequencing can improve confidence that an organism is, in fact, from the contaminant pool (compare, for example, Fig. 2.1A and C). However, even when sequencing depth is substantial, and even when contamination levels are low (e.g. 1000-fold less than input from the target environment), there is still a > 10% chance that the rarest OTU in a sequencing dataset is from the target community. Thus, whereas it is sometimes possible to reduce the likelihood of attributing contaminant organisms to the target environment with stringent control of contamination and strict thresholds on rank abundances, it is nearly impossible to capture all rare taxa from the target community unless one is willing to accept the inclusion of all contaminant OTUs.

In the analysis that we present, we have set a very strict limit on reliable OTUs. In particular, we have classed an OTU as reliable when there is a < 1% probability that the OTU is from the contaminant pool. Loosening this criterion, however, does not improve outcomes substantially. Indeed, if we accept a < 10% probability that an OTU is from the contaminant pool, then at low contamination ($w_{\text{contaminant}} = 0.001$ ng), a 200 OTU sample with equal partitioning of diversity between contamination and the target environment and no OTU overlap ($\theta = 0.5$, $\varphi = 0$) goes from having 29.5% reliable OTUs at the 1% threshold to having 33% reliable OTUs at the 10% threshold. For a comparable scenario at high contamination ($w_{\text{contaminant}} = 0.1$ ng), 3.5% of OTUs are reliable at the 1% threshold and 6% are reliable at the 10% threshold. Such modest gains are not unexpected given the steepness of the curves in Fig. 2.1, and are a direct result of the shape of the species abundance distribution with its long tail of rare organisms.

One major assumption of our model is that microbial populations in both the contaminant source and in the target environment are log-series distributed. Although many species abundance distributions, including those from microbial systems (Hill *et al.*, 2003) can be fit well to a log-series model, this is not always the case (Ulrich *et al.*, 2010; Hughes, 1986). Indeed, there may be communities where a log-normal distribution, or some other more complex model is better suited. Despite this, most species abundance distributions are likely to follow a general pattern of having a few abundant OTUs, followed by a large tail of rare OTUs; thus, many of the qualitative predictions of the current model should remain true, though precise quantitative results may deviate slightly if other distributions are used. Another assumption of our model is that all DNA comes from cells. That is, we use DNA quantification to arrive at total cell counts, which are then used to parameterize log-series species abundance distributions for both the contaminant pool and the target environment. In many sequencing studies, however, both the target environment and contamination sources can contribute free DNA. While UV-radiation can help to mitigate the input of free DNA to some extent, it remains an open question how much the presence of free DNA in sequencing samples could impact our predictions.

Contamination in sequencing studies remains one of the greatest obstacles to successful interpretation of sequencing data and, as a consequence, accurate characterization of microbial communities. In this work, we have taken a step towards defining reasonable bounds on OTU reliability based on rank abundance. Although this approach is akin to some existing efforts to remove contaminants based on a relative abundance threshold, our work cautions against such a method for several reasons. First, even for similar levels of contamination, the relative abundance threshold for trustworthy OTUs can change by at least an order of magnitude based on sample characteristics, most notably diversity and partitioning of OTU diversity between the sample and the contaminant source. Second, the relationship between contaminant input and a reliable relative abundance threshold is complex (see Fig. 2.5). Indeed, the nonlinearities we identified mean that one cannot merely scale the acceptable relative abundance threshold with estimated contamination levels. Rather, our model makes clear that optimal delineation between sample and contaminant OTUs requires knowledge of not only contaminant levels, but also OTU diversity coming in to the sample. When such information is available, it is relatively straightforward to define a threshold (based on either rank or relative abundance) for trustable OTUs. Unfortunately, this threshold may be quite strict, highlighting the importance of controlling contamination as much as possible.

Acknowledgements

This material is based on work supported by, or in part by, the US Army Research Laboratory and the US Army Research Office under contract/grant number #W911NF-14-1-0490.

Appendix 1

Assuming an average genome size of 4×10^6 bps, 1 ng of bacterial DNA will contain approximately 225,000 cells. Given the distribution in Vetrovsky *et al.* (2013), the average 16S copy number per cell is 4.17, meaning that there will be 938,250 16S sequences prior to amplification. Assuming that the sequenced variable region is $\approx$100 bps, this means that there are 9.38×10^7 bp prior to amplification. In order to arrive at 400-fold DNA amplification:

$$400 = \frac{910^{11} + 9.3810^{7}(1+\varepsilon)^{p}}{910^{11}} \tag{2.5}$$

which, for $p = 38$, yields $\varepsilon = 0.49$, or approximately 50% efficiency.

References

Bewick, S., Chisholm, R.A., Akçay, E., and Godsoe, W. (2015). A stochastic biodiversity model with overlapping niche structure. Theor. Ecol. *8*, 81–109.

Bewick, S., Thielen, P., Mehoke, T., Karig, D., and Fagan, W.F. (2017). Sampling, sequencing and the SAD. Ecol. Complex. *32*, 168–180.

Boswell, M., and Patil, G. (1971). Chance mechanisms generating the logarithmic series distribution used in the analysis of number of species and individuals. Statistical Ecology *1*, 99–130.

Champlot, S., Berthelot, C., Pruvost, M., Bennett, E.A., Grange, T., and Geigl, E.M. (2010). An efficient multistrategy DNA decontamination procedure of PCR reagents for hypersensitive PCR applications. PLOS ONE 5, e13042. https://doi.org/10.1371/journal.pone.0013042

Corless, C.E., Guiver, M., Borrow, R., Edwards-Jones, V., Kaczmarski, E.B., and Fox, A.J. (2000). Contamination and sensitivity issues with a real-time universal 16S rRNA PCR. J. Clin. Microbiol. *38*, 1747–1752.

Evans, G.E., Murdoch, D.R., Anderson, T.P., Potter, H.C., George, P.M., and Chambers, S.T. (2003). Contamination of Qiagen DNA extraction kits with Legionella DNA. J. Clin. Microbiol. *41*, 3452–3453.

Fisher, R.A., Corbet, A.S., and Williams, C.B. (1943).The relation between the number of species and the number of individuals in a random sample of an animal population. J. Animal Ecol. 42–58.

Fox, J.C., Ait-Khaled, M., Webster, A., and Emery, V.C. (1991). Eliminating PCR contamination: is UV irradiation the answer? J. Virol. Methods 33, 375–382.

Gill, P., Whitaker, J., Flaxman, C., Brown, N., and Buckleton, J. (2000). An investigation of the rigor of interpretation rules for STRs derived from less than 100 pg of DNA. Forensic Sci. Int. *112*, 17–40.

Gill, S.R., Pop, M., Deboy, R.T., Eckburg, P.B., Turnbaugh, P.J., Samuel, B.S., Gordon, J.I., Relman, D.A., Fraser-Liggett, C.M., and Nelson, K.E. (2006). Metagenomic analysis of the human distal gut microbiome. Science *312*, 1355–1359.

Grice, E.A., Kong, H.H., Conlan, S., Deming, C.B., Davis, J., Young, A.C., Bouffard, G.G., Blakesley, R.W., Murray, P.R., Green, E.D., *et al.* (2009). Topographical and temporal diversity of the human skin microbiome. Science *324*, 1190–1192. https://doi.org/10.1126/science.1171700

Hill, J.K., and Hamer, K.C. (1998). Using species abundance models as indicators of habitat disturbance in tropical forests. J. Applied Ecol. *35*, 458–460.

Hill, T.C., Walsh, K.A., Harris, J.A., and Moffett, B.F. (2003). Using ecological diversity measures with bacterial communities. FEMS Microbiol. Ecol. *43*, 1–11. https://doi.org/10.1111/j.1574-6941.2003.tb01040.x

Hughes, R. (1986). Theories and models of species abundance. Am. Nat. *128*, 879–899.

Kemp, A. (1981). Efficient generation of logarithmically distributed pseudo-random variables. App. Stat. J. Roy. St. C. 249–253.

Kempton, R., and Taylor, L. (1974). Log-series and log-normal parameters as diversity discriminants for the Lepidoptera. J. Animal Ecol. 381–399.

Kendall, D.G. (1948). On some modes of population growth leading to RA Fisher's logarithmic series distribution. Biometrika *35*, 6–15.

Lauder, A.P., Roche, A.M., Sherrill-Mix, S., Bailey, A., Laughlin, A.L., Bittinger, K., Leite, R., Elovitz, M.A., Parry, S., and Bushman, F.D. (2016). Comparison of placenta samples with contamination controls does not provide evidence for a distinct placenta microbiota. Microbiome *4*, 29.

Laurence, M., Hatzis, C., and Brash, D.E. (2014). Common contaminants in next-generation sequencing that hinder discovery of low-abundance microbes. PLOS ONE *9*, e97876. https://doi.org/10.1371/journal.pone.0097876

Lusk, R.W. (2014). Diverse and widespread contamination evident in the unmapped depths of high throughput sequencing data. PLOS ONE *9*, e110808. https://doi.org/10.1371/journal.pone.0110808

May, R.M. (1975). Patterns of species abundance and diversity. In Ecology and Evolution of Communities, Cody M.L., and Diamond J.M., eds (Harvard University Press, Cambridge, USA), pp. 81–120.

Salter, S.J., Cox, M.J., Turek, E.M., Calus, S.T., Cookson, W.O., Moffatt, M.F., Turner, P., Parkhill, J., Loman, N.J., and Walker, A.W. (2014). Reagent and laboratory contamination can critically impact sequence-based microbiome analyses. BMC Biol. *12*, 87. https://doi.org/10.1186/s12915-014-0087-z

Schmidt, T., Hummel, S., and Herrmann, B. (1995). Evidence of contamination in PCR laboratory disposables. Naturwissenschaften *82*, 423–431.

Ulrich, W., Ollik, M., and Ugland, I. (2010). A meta-analysis of species–abundance distributions. Oikos *119*, 1149–1155.

Větrovský, T., and Baldrian, P. (2013). The variability of the 16S rRNA gene in bacterial genomes and its consequences for bacterial community analyses. PLOS ONE *8*, e57923. https://doi.org/10.1371/journal.pone.0057923

Weiss, S., Amir, A., Hyde, E.R., Metcalf, J.L., Song, S.J., and Knight, R. (2014). Tracking down the sources of experimental contamination in microbiome studies. Genome Biol. *15*, 564. https://doi.org/10.1186/s13059-014-0564-2

Williamson, M., and Gaston, K.J. (2005). The lognormal distribution is not an appropriate null hypothesis for the species–abundance distribution. J. Animal Ecol. *74*, 409–422.

Molecular Methods to Study Microbial Succession in Soil

3

Francisco Dini-Andreote[1†], Xiu Jia[2†] and
Joana Falcão Salles[2*]

[1]Department of Microbial Ecology, Netherlands Institute of Ecology (NIOO-KNAW), Wageningen, the Netherlands.
[2]Microbial Ecology cluster, Genomics Research in Ecology and Evolution in Nature (GREEN), Groningen Institute for Evolutionary Life Sciences (GELIFES), University of Groningen, the Netherlands.

*Correspondence: j.falcao.salles@rug.nl

†Both authors contributed equally to this work

https://doi.org/10.21775/9781912530021.03

Abstract

The soil environment teems with microbes that play fundamental roles in controlling biogeochemical cycles, thus providing support for organisms above and below ground to thrive. With the contemporary development of high-throughput DNA sequencing technologies, a growing body of literature has shown that organismal composition and abundance within microbial communities can orderly and sequentially change through time. Interestingly, these patterns of ecological succession share commonalities across distinct systems, e.g. receding glacier forelands, salt marsh chronosequence, volcanic deposited gradients, post-mining areas and abandoned agricultural fields. Of critical importance, most of the available literature is based on community taxa distribution, with relatively little known about the successional changes in community functional traits. Here, we provide an overview of advances in molecular methods based on DNA sequencing and describe how these methods can enhance our understanding of the dynamics of microbial communities, with particular implications for studying patterns of microbial succession in soils. Collectively, the use of molecular methods can promote a comprehensive understanding of how microbial communities systematically change over both spatial and temporal scales, thus providing the basis for predicting microbial community responses after short- and long-term disturbances and in future environmental change scenarios.

The greatest diversity of life lies beneath our feet: a glimpse into the modern era of the molecular microbiology of soils

Microorganisms are the major determinants of the physical, chemical and biological characteristics of soil (van Elsas *et al.*, 2006; Prosser, 2015). It is well known among microbial ecologists that soils represent, by far, the most complex and diverse living ecosystem on earth (Whitman *et al.*, 1998). Perhaps one of the best geological definitions describes soil as being a highly complex environment of aggregated particles that create an intricate three-dimensional network of water- and air-filled pores (Oades, 1984). However, a missing link in this definition is the living fraction of soils. Soil is a complex physicochemical matrix that provides a myriad of niches to its inhabitants. A single gram of soil often contains about 10^9 bacterial individuals, encompassing 10^3–10^6 distinct taxa (Torsvik *et al.*, 2002, Gans *et al.*, 2005, Tringe *et al.* 2005), in addition to numerous other microorganisms (i.e. archaea, fungi, protists, nematodes and viruses) and metazoans. Interestingly, although soils teem with life, the fraction of the soil surface area that is covered by soil microbes is $<< 1\%$, which closely approximates the fraction of the surface area on earth that humans occupy (Young *et al.*, 2008). These soil-borne microorganisms play fundamental roles in a vast array of terrestrial ecosystem functions, which include the biogeochemical cycling of soil nutrients (nitrogen, carbon, sulfur, phosphorus, etc.) (Falkowski *et al.*, 2008), the sustenance of plant growth, water purification, carbon storage and the maintenance of soil physical structure (Young *et al.*, 2008; Vos *et al.*, 2013).

The study of microbes in soils has undergone a major breakthrough in the past two decades, with continuing advances in molecular sequencing technology circumventing methodological limitations. This has been coined the 'great plate count anomaly', in which only a small fraction of the bacterial cells in a soil sample (ca. 1–5%) is able to be successfully cultivated under standard laboratory conditions (Staley and Konopka, 1985). Together with direct nucleic acid isolation methods, continuous advances in DNA sequencing technologies have allowed the assessment of larger fractions of the soil microbial communities through the targeting of particular marker genes (e.g. the bacterial 16S rRNA gene, the fungal 18S rRNA gene or the internal transcribed spacer, ITS), that is, amplicon sequencing, used for community profiling, and the direct sequencing of DNA samples obtained from soils, that is, metagenomics.

It is important to notice that the use of high-throughput amplicon sequencing of soil samples extends beyond the spectrum of taxonomic markers (referred to above), being also applied to target functional genes, thus providing information on the genetic diversity of microbial functional groups (Pester *et al.*, 2012; Pereira e Silva *et al.*, 2013). In both cases, this strategy is used to provide information on the taxonomic/phylogenetic community composition and structure of microbial communities. The outcome information consists of an overview of the taxonomic/phylogenetic divergence of the groups present within and across samples, a set of ecological indices, that is, alpha-diversity (number of 'species' and the evenness of species within a community) and beta-diversity (metrics that quantify the degree to which species change across time, space and/or communities).

The term 'metagenomics' refers to the collection of genomes and genes of a given microbial community. Metagenomic information can be obtained through the direct high-throughput sequencing of the total DNA present in the samples (i.e. shotgun metagenomics). This strategy allows for the characterization of both phylogenetic and functional gene repertoires of microbial communities (Venter *et al.*, 2004). In the field of microbial ecology and biotechnology, metagenomics has arguably been revolutionizing our understanding of the composition and functional complexity of microbial communities during the past decades. The use of metagenomics is fundamental for the provision of novel genetic information, genomic linkages between function and phylogeny of uncultured microorganisms, and discoveries of novel functional genes, biocatalysts and enzymes. In addition, metagenomics can largely contribute to a better understanding of the ecological and evolutionary processes shaping microbial community dynamics across spatio-temporal scales in soils.

Considering the fact that methodological advances inevitably come with inherent limitations, the advances in other 'omics' technologies have been used in a complementary manner, leading to the so-called 'multi-omics information pipeline' (Lamendella *et al.*, 2012). In brief, this promising (yet complex) pipeline aims to integrate DNA profiling data with information about gene expression/activities (metatranscriptomics) (Shi *et al.*, 2009) and identification of protein profiles (metaproteomics) (Verberkmoes *et al.*, 2009), which collectively intends to corroborate the characterization of metabolites (metabolomics) (Bundy *et al.*, 2009). It is assumed that the joint application of these methods, in spite of the complexity and challenges that they pose, will provide a robust signature of community processes and activities in soil (Franzosa *et al.*, 2015).

Given the enormous progress in our ability to address microbial communities in soils, many outstanding questions in soil microbial ecology have started to be answered. For example, the taxon–area relationship has been found to be applicable to microbial communities (Horner-Devine *et al.*, 2004), and the greater understanding of microbial biogeographical patterns has provided evidence for the intuitive contention that microbes are often dispersal limited (Martiny *et al.*, 2006; Hanson *et al.*, 2012). Additionally, mounting evidence has shown that bacterial communities in soils are largely structured by differences in pH (Fierer and Jackson, 2006, Lauber *et al.*, 2009) and salinity (Lozupone and Knight, 2007), even across continental scales. Moreover, at local sites, a suite of other factors that create environmental variability can exert an influence on microbial populations across different soil systems. These include plant cover and productivity (Berg and Smalla, 2009), animal activity (Clegg, 2006), wetness (Schimel *et al.*, 1999), and fertilizer application (Fierer *et al.*, 2012). Using these few examples of early comprehensive studies, it is now possible to envisage that microbial ecologists face an era of unprecedented transformation. In particular, the integration of the growing knowledge of soil microbiology into the general field of ecology, such as the themes of biological diversity, invasion, community assembly, resistance, and resilience, is a very challenging and promising area. Now it is conceivable, through increasingly powerful survey tools, to develop high-resolution field assessments, creative experiments and ecological models that effectively contribute to the continuous use of microbial communities as model entities to study relevant questions in the discipline of community ecology. These developments represent a path towards a better understanding of the eco-evolutionary mechanisms that shape life in soil, thus leading to the new era of molecular soil microbiology.

Successional patterns of soil microbial communities

Ecological succession can be simply defined as how orderly, and in a systematic manner, organismal composition and abundance change over time (Fierer *et al.*, 2010). This concept was initially developed in the study of plant communities (Clements, 1916). The theme of ecological succession in soils comes from rather traditional concepts in ecology that are as yet not fully appreciated in soil microbiology. This is due, in part, to methodological impairments associated with observing microbes in nature and as a consequence of different historical paths taken by the disciplines of microbiology and general ecology (Jessup *et al.*, 2004; Prosser *et al.*, 2007). However, over the last two decades, microbial ecologists have witnessed unprecedented advances in their technical abilities to access microbial communities, in particular across a vast range of soil ecosystems worldwide [e.g. Terragenome Project Consortium (Vogel *et al.*, 2009), Earth Microbiome Project (Gilbert *et al.*, 2014)]. As a result, long-standing questions in soil microbiology, ranging from the classical 'who is where?' and 'what are they doing?' to more process-oriented questions such as 'how will they respond?', can now be answered on the basis of data obtained at high resolution from experimental or observational studies of the geographic and habitat distributions of microorganisms in living ecosystems (Martiny *et al.*, 2006; Jansson and Prosser, 2013). Currently, a growing number of studies have demonstrated that contemporary environmental conditions are able to explain the distribution of microorganisms (Hanson *et al.*, 2012). However, some researchers have suggested that historical conditions also contribute to microbial species assemblages (Fukami, 2015; Hawkes and Keitt, 2015). In this context, the study of soil microbial community succession provides an opportunity to overcome this conflict and resolve the mechanisms driving soil community structure in concert with contemporary and historical aspects.

The field of microbial ecological succession in soils is still recent, with the majority of the available literature relying on the investigation of community dynamics in receding glacier forelands (Sigler and Zeyer, 2002; Nemergut *et al.*, 2007; Schmidt *et al.*, 2008), post-mining areas (Hüttl and Weber, 2001; Kozdrój and van Elsas, 2001), abandoned agricultural fields (Elhottová *et al.*, 2002; Kuramae *et al.*, 2010) and forest fields that have experienced wildfire disturbances (Ferrenberg *et al.*, 2013). A common feature of many of these studies is the examination of the dynamism of the microbial communities across natural trajectories of landscape formation (primary succession) or recovery, in the case of disturbance (secondary succession) (Fig. 3.1). This occurs through the intertwined influence of shifting local abiotic factors and the biotic activity of soil microorganisms. Critical for such studies is the use of so-called chronosequences, as environmental models for temporal soil ecology (Box 3.1). The inherent assumption of space-per-time replacement of chronosequences allows the interrogation of ecological and potentially physiological changes in the communities across temporal scales – which in the case of microbes, can range from hours to months or decades to millennia – in a contemporary manner. This approach is based on the assumption that early successional stages change systematically in a way that is similar to the trajectories previously displayed by mature stages in the past (Johnson and Miyanishi, 2008; Walker *et al.*, 2010).

Ecological succession is conceptually divided into two categories, primary and secondary, with primary succession occurring in a non-colonized environment and secondary succession occurring in a previously colonized environment following a disturbance event (Fig. 3.1). In microbial ecology, some authors have proposed that such definition should be abandoned as the dynamic changes in microbial communities appear to be far more complex (Fierer *et al.*, 2010). As an alternative, these authors particularly propose that microbial succession can be envisaged by three general operationally defined categories based on carbon assimilation metabolism, all of which would traditionally be considered forms of primary

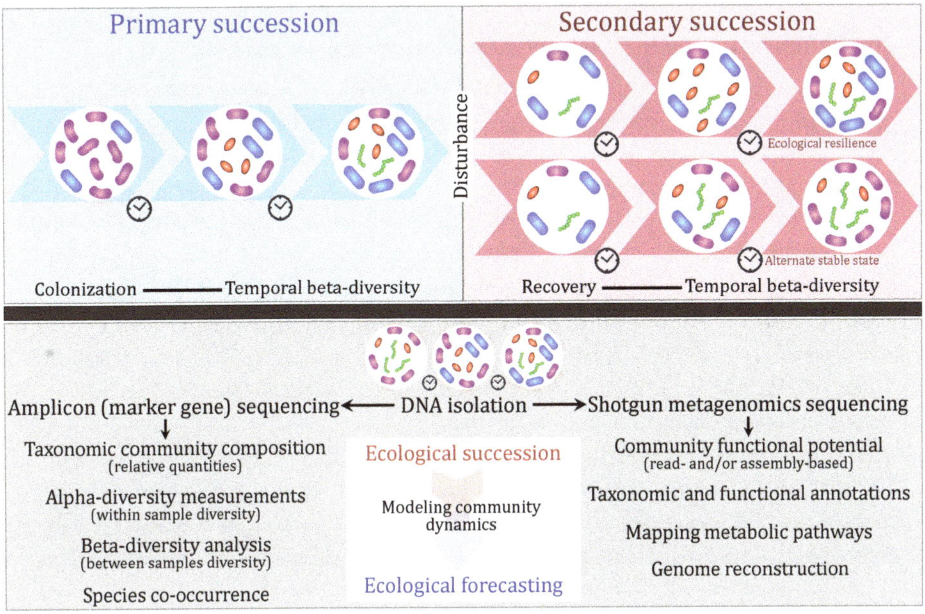

Figure 3.1 Schematic illustration depicting primary and secondary succession in microbial communities. The lower panel illustrates molecular methods based on DNA sequencing (amplicon sequencing and shotgun metagenomics) and their respective data outputs.

succession: (i) *autotrophic succession*: autotrophs are predominantly initial colonizers, that is, photoautotrophs or chemoautotrophs – the initial community develops slowly as the organic carbon supply in the system comes mostly from the initial colonizers themselves; (ii) *endogenous heterotrophic succession*: heterotrophs are predominantly initial colonizers that respire or ferment organic compounds to generate energy – the initial succession can be fast as it is fuelled by organic carbon derived from the available substrate, but, as succession progresses, the community development can slow down as the carbon resource is consumed and modified over time; (iii) *exogenous heterotrophic succession*: heterotrophic respires or fermenters are predominantly initial colonizers – the organic carbon is supplied constantly by external inputs, which can be highly variable over time; the ecological succession can also be fast in the initial stages but vary, in terms of turnover, at intermediate and late successional stages. However, given that this framework remains still to be properly tested and evaluated [see Fierer *et al.* (2010) for additional details], here we have used the more generalized concept based on colonization (primary succession) and recovery (secondary succession) to discuss the successional patterns of soil microbial communities.

Primary succession in soil microbial communities

Patterns of microbial primary succession in soils can be observed in distinct ecosystems, for example along the glacial till (Hell *et al.*, 2013), during the microbial colonization of an emerging island (Marteinsson *et al.*, 2015), after cycles of volcanic deposits (Weber and King, 2010) and in the rhizosphere of plants along the course of phenological development (Chaparro *et al.*, 2014). Despite idiosyncrasies among these systems, patterns of microbial succession share commonalities (albeit expectation may occur), for instance microbial

Box 3.1 The use of chronosequence to study temporal soil ecology

The study of temporal soil ecology is often challenged by setting appropriate timescales in a realistic and feasible analytical manner. This becomes even more complex when the investigated patterns occur over timescales that exceed decades or centuries. In these cases, the most appropriate method relies on the use of environmental chronosequences. The method is based on the study of established sites presumed to represent a sequence of different stages of soil development, with a known initial establishment and a time history afterward (Walker and Del Moral, 2003). As such, a fundamental assumption regarding chronosequences is that the communities (either micro or macro) and local environmental/ecosystem parameters of the younger soil sites are gradually developing in a temporal manner that resembles how the older sites developed (termed 'space-for-time substitution').

Well-known soil chronosequences are those formed by the worldwide recession of glaciers that have occurred over the past 150–250 years, leaving glacier forelands with spatially ordered sequences of terrain age (Bardgett *et al.*, 2005). In addition, spatially ordered sequences of terrain can also be found at abandoned fields over the course of several decades, at sand dunes covering centuries or in areas flanking volcanos, where temporally distinct events of lava flow promote the chronosequence development. Moreover, the established soil chronosequence was also found to concur with a natural salt marsh ecosystem formation (Olff *et al.*, 1997).

Finally, as chronosequences intrinsically imply the presence of ecological succession (Walker *et al.*, 2010), these model systems stand out as natural experiments that allow the testing of fundamental questions in temporal soil ecology. A few examples are (i) 'Do the same macro-ecological patterns apply to micro- and macro-organisms and are they caused by the same processes?; (ii) 'How do spatial and temporal environmental heterogeneities influence diversity at different scales?'; (iii) 'What are the relative roles of stochastic and deterministic processes controlling diversity and composition of communities?'; and, lastly, (iv) 'What are the most appropriate baselines for determining the magnitude and direction of ecological changes?' These four questions are far from being a complete list; they merely represent the key questions from the 'Identification of 100 fundamental ecological questions' (Sutherland *et al.*, 2013), in which chronosequences can be applied as a model system.

biomass tends to increase over time, as well as the complexity of biotic/abiotic conditions structuring the existing microbial communities. These patterns are mostly determined by orderly changes of abiotic variables in the soil and the continuous influence exerted by the dynamic establishment of plants in these systems. Often, the initial stages of succession present a higher increase in the rate of biomass accumulation in the soil, steadily stabilizing as succession proceeds. In addition, the temporal turnover of species (beta-diversity) is expected to be higher in the initial stages given the dynamics of the system, with a progressive decrease as the system matures (Dini-Andreote *et al.*, 2014).

Historically, studies on ecological succession have been carried out mainly in glacier ecosystems, which – because of the deglaciation process – offer a unique opportunity to investigate several decades of succession over a distance of a few hundred metres (Sigler *et al.*, 2002). Studies focusing on this ecosystem have already assessed changes in microbial community structures (Sigler *et al.*, 2002; Sigler and Zeyer, 2002; Nemergut *et al.*, 2007), functional diversity (Deiglmayr *et al.*, 2006; Kandeler *et al.*, 2006) and enzymatic activity (Tscherko *et al.*, 2003) over time. For instance, some studies on bacterial community composition have shown an increase in phylotype diversity over time following glacier retreat (Nemergut *et al.*, 2007), whereas others have found the opposite (Sigler *et al.*, 2002; Sigler and Zeyer, 2002). The latter findings suggest that the development of bacterial species richness may be the opposite of that observed in plant succession, where species richness increases over time (Godwin, 1929; Huston and Smith, 1987; Del Moral and Jones, 2002; Hodkinson *et al.*, 2003). However, the limited number of studies performed on microbial succession, and the different methodological approaches used to estimate microbial diversity, hinders any broad conclusions. In relation to community functions, some studies have described bacterial activity as presenting a transient increase at the initial successional stages followed by a decrease over time ('hump-back' pattern) (Schipper *et al.*, 2001; Sigler *et al.*, 2002), and microbial processes (e.g. nitrogen mineralization, ammonium oxidation, arginine deaminase) and soil enzyme activity (e.g. urease, protease, xylanase, phosphatase, arylsulphatase) were found to reach a steady state 50 years after glacier retreat (Tscherko *et al.*, 2003).

It should be noticed that, within the same system, bacterial and fungal communities may exhibit different successional trajectories during succession. For instance, by using amplicon sequencing, Brown and Jumpponen (2014) showed that bacterial communities converge more in terms of community similarity than fungi, along a glacier chronosequence. In a meta-analysis study, Zhou *et al.* (2017) illustrated that the fungal-to-bacteria ratio (*fungi/bacteria*) increased across successional stages. Specifically, *r*-strategists (bacteria) that had higher respiration per unit biomass carbon (R/C_{mic}) and lower microbial biomass carbon-to-soil carbon ratio (C_{mic}/C_{soil}) tend to be dominant in the early successional stages, whereas *K*-strategists (fungi) tend to prevail in more mature soil stages. In addition, bacteria exhibit a broader range of physiologies than fungi, for example bacteria can be photoautotrophs, heterotrophs or chemoautotrophs, whereas fungi are all heterotrophs. Thus, bacteria are more likely to successfully colonize early successional stages, where photo- and chemoautotrophs can thrive, whereas fungi, given their dependency on organic matter availability, tend to become more successful at later stages of succession (Schmidt *et al.*, 2014).

Secondary succession in soil microbial communities

Secondary succession is the process of the reestablishment of a community following an ecological disturbance event that partially eliminates species from the original community pool (Horn, 1974). Experimental assays are often performed to simulate ecological disturbances and to investigate community responses through time. For example, Jurburg *et al.* (2017b) applied heat-shock treatments in soil microcosms and investigated the post-disturbance dynamics of bacterial community during 50 days. In doing so, the authors conceptualized a three-stage model of community recovery: (i) the initial phase was characterized by the increase in the abundance of surviving taxa; (ii) in a second phase, there was a significant decrease in community turnover; and (iii) in a third phase the community reaches a 'stable' state that is similar to the initial structure prior to disturbance (ecological resilience) or not (alternative stable state). Notably, these authors described that phylogenetic turnover patterns showed a sharp increase in ecological selection in the initial

10 days after heat disturbance, which suggests strong competition in the initial stages of post-disturbance recovery. The role of biotic interactions driving the soil microbial dynamics in the early phase of secondary succession was in line with findings from Goberna *et al.* (2014), revealing that resource availability selects for competition-related traits (e.g. organic carbon consumption). In a follow-up experiment, Jurburg *et al.* (2017a) showed that microbial responses to disturbances are also dependent on the historical events experienced by a given community. For example, communities that experienced sequential perturbations (two cycles of heat-shock) tend to recover faster (or are more resilient) than those that experienced compounded perturbations (i.e. heat- and cold-shock) (Jurburg *et al.*, 2017a). The importance of historical events in determining community responses to disturbances was further confirmed in a microcosm experiment showing that the responses of soil bacterial communities to drought and flooding were dependent on land management (either sustainably managed or intensively grazed plots) (Jurburg *et al.*, 2018).

Plant facilitates soil microbial community succession

It is well known that, in soil, plant roots exert an influence on the assembly and structure of associated microbial communities. Of particular importance, the soil section tightly attached to the root and under its chemical influence is termed the rhizosphere. By comparing the successional trajectory of plants and microbial communities, Lozano *et al.* (2014) showed that above- and below-ground secondary succession were closely related. Plant roots provide one of the main primary sources of carbon that stabilize in the soil and then support heterotrophic microbes to thrive (Jones *et al.*, 2009). Using ^{14}C pulse labelling, Kuzyakov and Cheng (2001) showed that a substantial amount of carbon release in the rhizosphere was derived from the plant photo-assimilation. Therefore, it is unquestionable that successional patterns of plant-associated microbiomes are highly regulated by plant species, phenology, and physiology. For instance, by studying the root microbiome of 31 plant species that grow along a coastal tropical soil chronosequence that spans ca. 460,000 years, Yeoh *et al.* (2017) found that host phylogeny together with soil types affect the root-associated microbial community composition during ecological succession. In addition, by combining stable isotope probing with phospholipid analysis and amplicon-sequencing analyses, Hannula *et al.* (2017) showed changes in the active plant-associated microbiome along a secondary successional gradient of abandoned agricultural fields. In particular, these authors found that root-derived carbon was largely metabolized by bacteria in recently recovering fields, whereas at later stages these resources were mostly utilized by fungi. It was shown that, within the fungal community, the dominance of fast-growing and pathogenic species was progressively replaced by slow-growing and beneficial fungal species through time. This pattern indicates a later contribution of fungi to plant–soil feedback, where fungi play a role in soil nutrient cycling and assist plant community development during secondary succession.

The functional perspective of soil microbial community succession

Patterns of microbial community succession can also be investigated in light of community traits and functions. Such a strategy provides a link between biodiversity patterns and ecosystem functioning (Reiss *et al.*, 2009). For example, using an ancestral trait reconstruction approach (i.e. PICRUST; Langille *et al.*, 2013), Nemergut *et al.* (2016) showed that the average rRNA copy numbers significantly declined from more than 9 copies in the early successional stages to ca. 3.6 copies in the late stages. In general, organisms containing more copies of this gene are expected to more efficiently metabolize nutrients and quickly respond to nutrient inputs (Klappenbach *et al.*, 2000; Stevenson and Schmidt, 2004; Yano *et al.*, 2013). It implies that opportunistic, fast-growing *r*-strategists are more adapted to early successional stages, whereas slow-growing *K*-strategists are likely to be more successful in late soil succession stages. These authors further suggested that the consistency of this finding is a result of strong ecological selection, as evidenced by a meta-analysis encompassing a set of other successional gradients, including a salt marsh chronosequence (Dini-Andreote *et al.*, 2015) and a post-fire disturbance soil system (Ferrenberg *et al.*, 2013).

Other functional aspects of microbial communities also vary in response to soil colonization by microbes during primary and secondary succession, although the number of studies investigating these changes is still relatively scarce. For instance, Cline and Zak (2015) recently showed that the input of plant detritus, which are known to vary along succession, has an important influence on the microbial community structure, in particular, fungal communities. By applying amplicon sequencing to probe fungal and bacterial community compositions and shotgun metagenomics to assess community functional traits, these authors showed that plants exert a significant influence on the turnover of soil fungal communities, whereas edaphic properties (i.e. pH) explain variations in bacterial communities. Specifically, the increasing abundance of lignin-rich C4 grasses leads to an increase in the activity of extracellular enzymes with lignocellulolytic activity, which was correlated with the composition of fungal genes associated with lignocellulolytic functions. In another study, Ortiz-Alvarez *et al.* (2018) performed a meta-analysis of microbial ecological succession and revealed that traits associated with nutrient cycling (i.e. nitrogen and carbon fixation, inorganic phosphate uptake and mobilization) systematically decrease as succession in soil spans. Collectively this suggests that selection operates along succession by favouring taxa adapted to less liable (i.e. more recalcitrant) resource conditions.

A case study: microbial community succession along a salt marsh chronosequence

The salt marsh located at the island of Schiermonnikoog (N53°30′ E6°10′), the Netherlands, offers a unique opportunity to study ecological succession along a soil chronosequence. This is a well-documented ecosystem gradient spanning more than a century of succession (Olff *et al.*, 1997). The island system continuously extends eastwards, as sediments are deposited by wind and sea currents from west to east (Dini Andreote, 2016) (Fig. 3.2).

The early stages are formed on the east side of the island, under the constant influence of the tides, causing daily aerobic–anaerobic shifts and variation in temperature and moisture. As succession spans, there is a progressive increase in soil nutrient concentration and a decrease in the variability of environmental conditions. The later stages of the succession are less subjected to tidal influence and display more structured soils and abundant vegetation.

In this system, soil physicochemical parameters vary significantly across the distinct successional stages. With succession, the sand content progressively declines from 92.3% at the initial stage to 13.3% at the late stage, whereas the silt and clay accrete over time (ranging from initially 2.7% silt and 5.0% clay at the early stage up to 49.0% and 37.7% at late stages). The amount of total soil organic matter increases along the chronosequence from 1 g/dm^3 at the initial stage to 31 g/dm^3 at the late stage. The pH decreases from 8.7 at early stages to

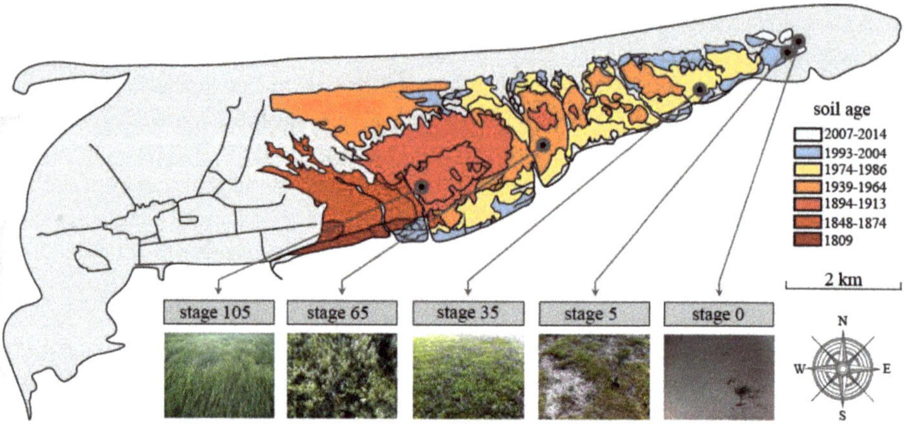

Figure 3.2 The salt marsh chronosequence located at the island of Schiermonnikoog, the Netherlands [figure adapted from Dini-Andreote *et al.* (2016b)]. The age of each stage of succession along the chronosequence was estimated from topographic maps, aerial photographs and the thickness of the sediment layer accumulating on the top of the underlying sand layer (Olff *et al.*, 1997; Schrama *et al.*, 2012).

7.4 at the late stages, whereas water content and the amount of sulfate, sodium, ammonium, nitrate and total nitrogen increase along the chronosequence (average value ranging from initially 12% water content, 49 mg/cm^3 sulfate, 443 mg/cm^3 sodium, 4 mg/ kg ammonium, 2 mg/kg nitrite and 385 mg/kg total nitrogen in the early stages up to 38%, 469 mg/cm^3, 4605 mg/cm^3, 68 mg/kg, 38 mg/kg and 5728 m/kg in the late stages, respectively) (Dini-Andreote *et al.*, 2014).

Former studies along this chronosequence have shown that macro- and microbial communities shift systematically along the system, each one following a specific distributional pattern. Regarding macro-organisms, the initial soil stages have low plant biomass (ca. 200 g/m2) covering only 8% of the soil, whereas the intermediate and late sites are more completely covered by vegetation (ca. 80% of the soil surface, biomass of ca. 600 g/m^2) (Schrama *et al.*, 2012). The composition of the vegetation changes along the chronosequence, from a *Plantago maritima*-dominated system in the initial stages, towards an *Artemisia maritima*-dominated at the intermediate stages, to finally later stages being vegetated by *Elytrigia atherica* (Olff *et al.*, 1997; Schrama *et al.*, 2012). By considering the flow of energy, Schrama *et al.* (2012, 2013) found that the initial stages of this chronosequence are largely under the influence of external marine-derived nutrients (i.e. the 'green food web'), whereas, as succession proceeds, soils are gradually transformed into a terrestrial environment, where the dynamics of the food web are driven by internal plant production (i.e. the 'brown food web').

With respect to microorganisms, Dini-Andreote *et al.* (2014, 2016b) found the abundance of soil bacterial and fungal communities to increase along this chronosequence. However, there were no significant related patterns of α-diversity of bacterial and fungal communities across the successional stages, which indicated that the mechanisms controlling species richness within these two organismal groups in soil may be different (Fig. 3.3) (Dini-Andreote *et al.*, 2014, 2016b).

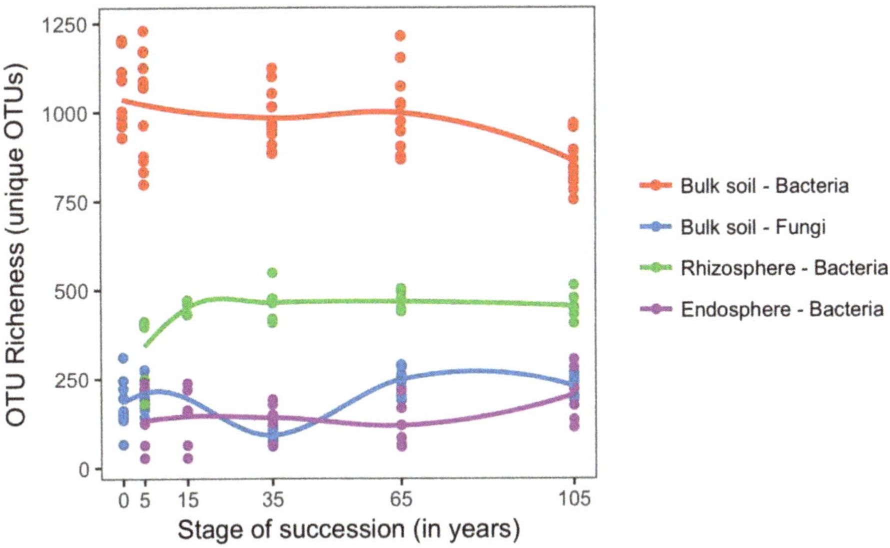

Figure 3.3 Soil microbial diversity, expressed as number of unique operational taxonomic units (OTUs), along a gradient of primary succession. Bacterial diversity is highest in the bulk soil (red), decreasing in diversity as the degree of association with the roots of *L. vulgare* increases (rhizosphere, green; endosphere, purple). Fungal diversity in the bulk soil (blue) is comparable to the diversity of endosphere-associated bacteria. The diversity of rhizosphere-associated bacteria follows the successional patterns associated with plant biomass (Schrama *et al.*, 2012; lowest at early stages and reaching a plateau at intermediate stages of succession) whereas the diversity of bacteria in the bulk soil is highest in the early stages, decreasing towards the end of the succession. Data adapted from Dini-Andreote *et al.* (2014), Dini-Andreote (2016) and Wang *et al.* (2015).

Specifically, for the bacterial community, there is a high turnover of bacterial communities at the initial stages, which indicates the role of temporally driven niche partitioning and stochasticity driving the initial assembly of bacterial communities in this chronosequence. As succession proceeds, the importance of deterministic selection progressively increases (i.e. environmental filtering becomes stronger), which was mostly attributed to the increasing concentration of sodium and variation in soil organic matter towards late successional stages (Dini-Andreote *et al.*, 2015). Like the bacterial community, fungal communities exhibited significant temporal variations at newly formed sites, which may be due to the dynamic conditions imposed by the tidal regime. Despite the lack of a common pattern between bacterial and fungal α-diversity, the β-diversity patterns were rather similar. The higher similarity observed at the two latest stages of succession for both bacterial and fungal community compositions was associated with the stability of soil physical structure and organic matter (Dini-Andreote *et al.*, 2016b).

Similar to the free-living bulk soil microbiome, bacterial communities associated with the rhizosphere of the plant *Limonium vulgare* – a common plant species that colonizes several stages of this chronosequence – were found to be distinct across succession stages. In addition, the endosphere-associated microbiome of *L. vulgare* (composed of microorganisms living inside the plant) differs from the bulk soil and rhizosphere microbiomes (Wang *et al.*, 2015). In particular, the endophytic bacterial communities were strongly associated with the selection imposed by the plant, and not clustering according to soil successional stages.

It can be concluded that, particularly in this salt marsh chronosequence, the tidal regime contributes significantly to the dispersal and initial colonization/establishment of both bacterial and fungal communities in soil. As the soil succession proceeds, the multiplicity of biotic and abiotic constraints progressively increases, leading to a soil microbial community being structured by a dynamic interplay of salinity and organic matter (content and quality). In the case of plant-associated microbiomes, it may be envisaged that such local selective pressures are alleviated in the rhizosphere, and likely to be inexistent in the endosphere.

Metagenomic analyses (shotgun metagenomics) of bulk soil samples revealed that the functional β-diversity patterns based on KEGG [Kyoto Encyclopedia of Genes and Genomes] Orthology [KO] were similar to those observed using the phylogenetic approach (amplicon sequencing based on the bacterial 16S rRNA gene and fungal ITS region), indicating congruence of both function and taxonomy (Dini-Andreote *et al.*, 2016a). Further examination of KOs associated with antibiotic resistance, motility and carbon metabolism revealed two distinct scenarios of community traits along the chronosequence, named the 'colonization' and 'competition' response modus. The 'colonization' modus – represented by the *in silico* reconstruction of the bacterial chemotaxis and flagellar assembly pathways – were traits that prevailed in microbiomes from the early successional stages as a reflection of the high diffusibility (i.e. connectivity) in these habitats. Under such conditions, motile chemosensory behaviour constitutes a key physiological asset that allows bacterial cells to perceive and exploit diverse micro-scale nutrient patches in a dynamic manner. On the other hand, in the more terrestrial (less connected) habitats, diffusibility becomes limited and chemically mediated mechanisms of biotic interaction become progressively more dominant, thus the importance of antibiotic resistance genes at the later stages of succession, the 'competition' modus. Moreover, both the number of genes associated with the degradation of carbon compounds as well as the complexity of the compounds increases towards late stages of succession in this chronosequence (Dini-Andreote *et al.*, 2018).

By focusing on the KOs associated with nitrogen cycling, Dini-Andreote *et al.* (2016a) showed that genes associated with nitrification (*amo* genes, coding for ammonia monooxygenase) were dominant in the first two stages of succession (peaking at 5 years), whereas KOs associated with denitrification (*nos*Z, *nor*B, and *nor*C genes) and biological nitrogen fixation (*nif*HDK) were dominant at the intermediate stage of succession (35 years). The quantification of specific genes and functional groups by quantitative polymerase chain reaction (qPCR) revealed that the distribution of nitrogen-cycling genes along the soil successional stages was driven by niche partitioning and abiotic variables. For instance, regarding the organisms involved in the first step of the nitrification processes [ammonia-oxidizing bacteria (AOB) and ammonia-oxidizing archaea (AOA)], it was shown that there is a dominance of AOB over AOA in this system. Moreover, the relative abundance of AOB was higher at the intermediate stage, whereas AOA abundance increased along succession (Dini-Andreote *et al.*, 2016), the latter being positively correlated with potential nitrification activities (Salles *et al.*, 2017). The relative abundance of *nir*K and *nir*S genes, which encode enzymes that are involved in nitrite reduction during denitrification, also vary in a distinct manner along the successional gradient, in which *nir*K peaked at the initial soil stages and steadily decreased over time, whereas *nir*S showed the opposite patterns (Dini-Andreote *et* *al.*, 2016a). Similarly, opposite patterns were observed for two *nosZ* clades (encode nitrous oxide reductase), with *nosZ* clade I being about 10-fold higher in abundance than the clade II (Dini-Andreote *et al.*, 2016a). The peak of *nosZ* clade II at the early stages of succession indicates that N_2O emission might be lower at this stage. In summary, these results illustrate the potential use of combinatory approaches (amplicon sequencing, metagenomics and qPCR) to probe and reconstruct the taxonomic and functional aspects of microbial communities in a dynamically evolving ecosystem.

Molecular methods in the context of soil microbial succession: future perspectives

Community ecologists who study plants and animals conceptually face a delineated set of emergent properties that are intrinsic to complex communities, namely biological diversity, functional redundancy and system stability. These are key general themes that serve as underlying properties from which more refined ecological questions can be raised. In microbial systems, as proposed by Konopka (2009), because microbes possess mechanisms for the horizontal transfer of genetic information and have a distinguished *modus operandi* (as we often study the system at a larger scale than that at which biological interactions occurs, because of analytical bottlenecks), the molecular methods may be considered as an emergent community property. As such, the use of community genetic information (i.e. the metagenome) constitutes a valuable piece of information to interrogate potential ecological traits as determinants of the successful establishment and dynamics of microbial communities. In line with this, soil metagenomes can be used as a property to access dynamic shifts in community traits. This can be performed at the level of metabolic redundancy and niche differentiation.

Finally, natural microbial communities are compositionally not stable, meaning that they are continuously in a flux (Shade *et al.*, 2013). This dynamic operates at different levels, from the micro- to the macro-scale, and it varies in space and across community types. For example, whereas at a micro-scale 'endogenous' dynamics in the relative abundance of community members and their patterns of gene expression operates, at a larger spatial scale, 'exogenous' influences exerted by shifting environmental factors prevail. As a result, a clear delineation of what may constitute the proper spatial and temporal scale to interrogate microbial systems is still a debatable topic (e.g. see Vos *et al.*, 2013). Thus, prospective frameworks that take into account not only the microbial fraction but also the whole ecosystem succession should be encouraged in the future.

References

Bardgett, R.D., Bowman, W.D., Kaufmann, R., and Schmidt, S.K. (2005). A temporal approach to linking aboveground and belowground ecology. Trends Ecol. Evol. *20*, 634–641.

Berg, G., and Smalla, K. (2009). Plant species and soil type cooperatively shape the structure and function of microbial communities in the rhizosphere. FEMS Microbiol. Ecol. *68*, 1–13. https://doi.org/10.1111/j.1574-6941.2009.00654.x

Brown, S.P., and Jumpponen, A. (2014). Contrasting primary successional trajectories of fungi and bacteria in retreating glacier soils. Mol. Ecol. *23*, 481–497. https://doi.org/10.1111/mec.12487

Bundy, J.G., Davey, M.P., and Viant, M.R. (2009). Environmental metabolomics: a critical review and future perspectives. Metabolomics *5*, 3.

Chaparro, J.M., Badri, D.V., and Vivanco, J.M. (2014). Rhizosphere microbiome assemblage is affected by plant development. ISME J. *8*, 790–803. https://doi.org/10.1038/ismej.2013.196

Clegg, C.D. (2006). Impact of cattle grazing and inorganic fertiliser additions to managed grasslands on the microbial community composition of soils. Appl. Soil Ecol. *31*, 73–82.

Clements, F.E. (1916). Plant succession: an analysis of the development of vegetation. Carnegie Institution of Washington.

Cline, L.C., and Zak, D.R. (2015). Soil microbial communities are shaped by plant-driven changes in resource availability during secondary succession. Ecology *96*, 3374–3385.

Deiglmayr, K., Philippot, L., Tscherko, D., and Kandeler, E. (2006). Microbial succession of nitrate-reducing bacteria in the rhizosphere of *Poa alpina* across a glacier foreland in the Central Alps. Environ. Microbiol. *8*, 1600–1612.

Del Moral, R., and Jones, C. (2002). Vegetation development on pumice at Mount St. Helens, USA. Plant Ecol. *162*, 9–22.

Dini-Andreote, F. (2016). Microbial community assembly in an evolving ecosystem: Ecological succession and functional properties of soil microbes. University of Groningen, Groningen.

Dini-Andreote, F., de Cássia Pereira e Silva, M., Triadó-Margarit, X., Casamayor, E.O., van Elsas, J.D., and Salles, J.F. (2014). Dynamics of bacterial community succession in a salt marsh chronosequence: evidences for temporal niche partitioning. ISME J. *8*, 1989–2001. https://doi.org/10.1038/ismej.2014.54

Dini-Andreote, F., Stegen, J.C., van Elsas, J.D., and Salles, J.F. (2015). Disentangling mechanisms that mediate the balance between stochastic and deterministic processes in microbial succession. Proc. Natl. Acad. Sci. U.S.A. *112*, E1326–32. https://doi.org/10.1073/pnas.1414261112

Dini-Andreote, F., Brossi, M.J., van Elsas, J.D., and Salles, J.F. (2016a). Reconstructing the genetic potential of the microbially-mediated nitrogen cycle in a salt marsh ecosystem. Front. Microbiol. *7*, 902. https://doi.org/10.3389/fmicb.2016.00902

Dini-Andreote, F., Pylro, V.S., Baldrian, P., van Elsas, J.D., and Salles, J.F. (2016b). Ecological succession reveals potential signatures of marine-terrestrial transition in salt marsh fungal communities. ISME J. *10*, 1984–1997. https://doi.org/10.1038/ismej.2015.254

Dini-Andreote, F. van Elsas, J.D., Olff, H., and Salles, J.F. (2018). Dispersal–competition tradeoff in microbiomes in the quest for land colonization. Sci. Rep. *8*, 9451.

Elhottová, D., Szili-Kovács, T., and Tríska, J. (2002). Soil microbial community of abandoned sand fields. Folia Microbiol. *47*, 435–440.

Falkowski, P.G., Fenchel, T., and Delong, E.F. (2008). The microbial engines that drive Earth's biogeochemical cycles. Science *320*, 1034–1039. https://doi.org/10.1126/science.1153213

Ferrenberg, S., O'Neill, S.P., Knelman, J.E., Todd, B., Duggan, S., Bradley, D., Robinson, T., Schmidt, S.K., Townsend, A.R., Williams, M.W., *et al.* (2013). Changes in assembly processes in soil bacterial communities following a wildfire disturbance. ISME J. *7*, 1102–1111. https://doi.org/10.1038/ismej.2013.11

Fierer, N., and Jackson, R.B. (2006). The diversity and biogeography of soil bacterial communities. Proc. Natl. Acad. Sci. U.S.A. *103*, 626–631.

Fierer, N., Nemergut, D., Knight, R., and Craine, J.M. (2010). Changes through time: integrating microorganisms into the study of succession. Res. Microbiol. *161*, 635–642. https://doi.org/10.1016/j.resmic.2010.06.002

Fierer, N., Lauber, C.L., Ramirez, K.S., Zaneveld, J., Bradford, M.A., and Knight, R. (2012). Comparative metagenomic, phylogenetic and physiological analyses of soil microbial communities across nitrogen gradients. ISME J. *6*, 1007–1017. https://doi.org/10.1038/ismej.2011.159

Franzosa, E.A., Hsu, T., Sirota-Madi, A., Shafquat, A., Abu-Ali, G., Morgan, X.C., and Huttenhower, C. (2015). Sequencing and beyond: integrating molecular 'omics' for microbial community profiling. Nat. Rev. Microbiol. *13*, 360–372. https://doi.org/10.1038/nrmicro3451

Fukami, T. (2015). Historical contingency in community assembly: integrating niches, species pools, and priority effects. Ann. Rev. Ecol. Evol. Syst. *46*, 1–23.

Gans, J., Wolinsky, M., and Dunbar, J. (2005). Computational improvements reveal great bacterial diversity and high metal toxicity in soil. Science *309*, 1387–1390.

Gilbert, J.A., Jansson, J.K., and Knight, R. (2014). The Earth Microbiome project: successes and aspirations. BMC Biol. *12*, 69. https://doi.org/10.1186/s12915-014-0069-1

Goberna, M., Navarro-Cano, J.A., Valiente-Banuet, A., García, C., and Verdú, M. (2014). Abiotic stress tolerance and competition-related traits underlie phylogenetic clustering in soil bacterial communities. Ecol. Lett. *17*, 1191–1201. https://doi.org/10.1111/ele.12341

Godwin, H. (1929). The sub-climax and deflected succession. J. Ecol. *17*, 144–147.

Hannula, S.E., Morriën, E., de Hollander, M., van der Putten, W.H., van Veen, J.A., and de Boer, W. (2017). Shifts in rhizosphere fungal community during secondary succession following abandonment from agriculture. ISME J. *11*, 2294–2304. https://doi.org/10.1038/ismej.2017.90

Hanson, C.A., Fuhrman, J.A., Horner-Devine, M.C., and Martiny, J.B. (2012). Beyond biogeographic patterns: processes shaping the microbial landscape. Nat. Rev. Microbiol. *10*, 497–506. https://doi.org/10.1038/nrmicro2795

Hawkes, C.V., and Keitt, T.H. (2015). Resilience vs. historical contingency in microbial responses to environmental change. Ecol. Lett. *18*, 612–625. https://doi.org/10.1111/ele.12451

Hell, K., Edwards, A., Zarsky, J., Podmirseg, S.M., Girdwood, S., Pachebat, J.A., Insam, H., and Sattler, B. (2013). The dynamic bacterial communities of a melting High Arctic glacier snowpack. ISME J. *7*, 1814–1826. https://doi.org/10.1038/ismej.2013.51

Hodkinson, I.D., Coulson, S.J., and Webb, N.R. (2003). Community assembly along proglacial chronosequences in the high Arctic: vegetation and soil development in north-west Svalbard. J. Ecol. *91*, 651–663.

Horn, H.S. (1974). The ecology of secondary succession. Ann. Rev. Ecol. Syst. *5*, 25–37.

Horner-Devine, M.C., Lage, M., Hughes, J.B., and Bohannan, B.J. (2004). A taxa-area relationship for bacteria. Nature *432*, 750–753.

Huston, M., and Smith, T. (1987). Plant succession: life history and competition. Am. Nat. *130*, 168–198.

Hüttl, R.F., and Weber, E. (2001). Forest ecosystem development in post-mining landscapes: a case study of the Lusatian lignite district. Naturwissenschaften *88*, 322–329.

Jansson, J.K., and Prosser, J.I. (2013). Microbiology: The life beneath our feet. Nature *494*, 40–41. https://doi.org/10.1038/494040a

Jessup, C.M., Kassen, R., Forde, S.E., Kerr, B., Buckling, A., Rainey, P.B., and Bohannan, B.J. (2004). Big questions, small worlds: microbial model systems in ecology. Trends Ecol. Evol. *19*, 189–197.

Johnson, E.A., and Miyanishi, K. (2008). Testing the assumptions of chronosequences in succession. Ecol. Lett. *11*, 419–431. https://doi.org/10.1111/j.1461-0248.2008.01173.x

Jones, D.L., Nguyen, C., and Finlay, R.D. (2009). Carbon flow in the rhizosphere: carbon trading at the soil–root interface. Plant Soil *321*, 5–33.

Jurburg, S.D., Nunes, I., Brejnrod, A., Jacquiod, S., Priemé, A., Sørensen, S.J., Van Elsas, J.D., and Salles, J.F. (2017a). Legacy effects on the recovery of soil bacterial communities from extreme temperature perturbation. Front. Microbiol. *8*, 1832. https://doi.org/10.3389/fmicb.2017.01832

Jurburg, S.D., Nunes, I., Stegen, J.C., Le Roux, X., Priemé, A., Sørensen, S.J., and Salles, J.F. (2017b). Autogenic succession and deterministic recovery following disturbance in soil bacterial communities. Sci. Rep. *7*, 45691. https://doi.org/10.1038/srep45691

Jurburg, S.D., Natal-da-Luz, T., Raimundo, J., Morais, P.V., Sousa, J.P., van Elsas, J.D., and Salles, J.F. (2018). Bacterial communities in soil become sensitive to drought under intensive grazing. Sci. Total Environ. *618*, 1638–1646.

Kandeler, E., Deiglmayr, K., Tscherko, D., Bru, D., and Philippot, L. (2006). Abundance of narG, nirS, nirK, and nosZ genes of denitrifying bacteria during primary successions of a glacier foreland. Appl. Environ. Microbiol. *72*, 5957–5962.

Klappenbach, J.A., Dunbar, J.M., and Schmidt, T.M. (2000). rRNA operon copy number reflects ecological strategies of bacteria. Appl. Environ. Microbiol. *66*, 1328–1333.

Konopka, A. (2009). What is microbial community ecology? ISME J. *3*, 1223–1230. https://doi.org/10.1038/ismej.2009.88

Kozdrój, J., and van Elsas, J.D. (2001). Structural diversity of microbial communities in arable soils of a heavily industrialised area determined by PCR-DGGE fingerprinting and FAME profiling. Appl. Soil Ecol. *17*, 31–42.

Kuramae, E.E., Gamper, H.A., Yergeau, E., Piceno, Y.M., Brodie, E.L., Desantis, T.Z., Andersen, G.L., van Veen, J.A., and Kowalchuk, G.A. (2010). Microbial secondary succession in a chronosequence of chalk grasslands. ISME J. *4*, 711–715. https://doi.org/10.1038/ismej.2010.11

Kuzyakov, Y., and Cheng, W. (2001). Photosynthesis controls of rhizosphere respiration and organic matter decomposition. Soil Biol. Biochem. *33*, 1915–1925.

Lamendella, R., VerBerkmoes, N., and Jansson, J.K. (2012). 'Omics' of the mammalian gut – new insights into function. Curr. Opin. Biotechnol. *23*, 491–500. https://doi.org/10.1016/j.copbio.2012.01.016

Langille, M.G., Zaneveld, J., Caporaso, J.G., McDonald, D., Knights, D., Reyes, J.A., Clemente, J.C., Burkepile, D.E., Vega Thurber, R.L., Knight, R., *et al.* (2013). Predictive functional profiling of microbial communities using 16S rRNA marker gene sequences. Nat. Biotechnol. *31*, 814–821. https://doi.org/10.1038/nbt.2676

Lauber, C.L., Hamady, M., Knight, R., and Fierer, N. (2009). Pyrosequencing-based assessment of soil pH as a predictor of soil bacterial community structure at the continental scale. Appl. Environ. Microbiol. *75*, 5111–5120. https://doi.org/10.1128/AEM.00335-09

Lozano, Y.M., Hortal, S., Armas, C., and Pugnaire, F.I. (2014). Interactions among soil, plants, and microorganisms drive secondary succession in a dry environment. Soil Biol. Biochem. *78*, 298–306.

Lozupone, C.A., and Knight, R. (2007). Global patterns in bacterial diversity. Proc. Natl. Acad. Sci. U.S.A. *104*, 11436–11440.

Marteinsson, V., Klonowski, A., Reynisson, E., Vannier, P., Sigurdsson, B.D., and Ólafsson, M. (2015). Microbial colonization in diverse surface soil types in Surtsey and diversity analysis of its subsurface microbiota. Biogeosciences *12*, 1191–1203.

Martiny, J.B., Bohannan, B.J., Brown, J.H., Colwell, R.K., Fuhrman, J.A., Green, J.L., Horner-Devine, M.C., Kane, M., Krumins, J.A., Kuske, C.R., *et al.* (2006). Microbial biogeography: putting microorganisms on the map. Nat. Rev. Microbiol. *4*, 102–112.

Nemergut, D.R., Anderson, S.P., Cleveland, C.C., Martin, A.P., Miller, A.E., Seimon, A., and Schmidt, S.K. (2007). Microbial community succession in an unvegetated, recently deglaciated soil. Microb. Ecol. *53*, 110–122. https://doi.org/10.1007/s00248-006-9144-7

Nemergut, D.R., Knelman, J.E., Ferrenberg, S., Bilinski, T., Melbourne, B., Jiang, L., Violle, C., Darcy, J.L., Prest, T., Schmidt, S.K., *et al.* (2016). Decreases in average bacterial community rRNA operon copy number during succession. ISME J. *10*, 1147–1156. https://doi.org/10.1038/ismej.2015.191

Oades, J.M. (1984). Soil organic matter and structural stability: mechanisms and implications for management. Plant Soil 76,319-337.

Olff, H., De Leeuw, J., Bakker, J.P., Platerink, R.J., and Van Wijnen, H.J. (1997). Vegetation succession and herbivory in a salt marsh: changes induced by sea level rise and silt deposition along an elevational gradient. J. Ecol. *85*, 799-814.

Ortiz-Álvarez, R., Fierer, N., de Los Ríos, A., Casamayor, E.O., and Barberán, A. (2018). Consistent changes in the taxonomic structure and functional attributes of bacterial communities during primary succession. ISME J. *12*, 1658–1667. https://doi.org/10.1038/s41396-018-0076-2

Pereira e Silva, M.C., Schloter-Hai, B., Schloter, M., van Elsas, J.D., and Salles, J.F. (2013). Temporal dynamics of abundance and composition of nitrogen-fixing communities across agricultural soils. PLOS ONE *8*, e74500. https://doi.org/10.1371/journal.pone.0074500

Pester, M., Rattei, T., Flechl, S., Gröngröft, A., Richter, A., Overmann, J., Reinhold-Hurek, B., Loy, A., and Wagner, M. (2012). amoA-based consensus phylogeny of ammonia-oxidizing archaea and deep sequencing of amoA genes from soils of four different geographic regions. Environ. Microbiol. *14*, 525–539. https://doi.org/10.1111/j.1462-2920.2011.02666.x

Prosser, J.I. (2015). Dispersing misconceptions and identifying opportunities for the use of 'omics' in soil microbial ecology. Nat. Rev. Microbiol. *13*, 439–446. https://doi.org/10.1038/nrmicro3468

Prosser, J.I., Bohannan, B.J., Curtis, T.P., Ellis, R.J., Firestone, M.K., Freckleton, R.P., Green, J.L., Green, L.E., Killham, K., Lennon, J.J., *et al.* (2007). The role of ecological theory in microbial ecology. Nat. Rev. Microbiol. *5*, 384–392.

Reiss, J., Bridle, J.R., Montoya, J.M., and Woodward, G. (2009). Emerging horizons in biodiversity and ecosystem functioning research. Trends Ecol. Evol. *24*, 505–514. https://doi.org/10.1016/j.tree.2009.03.018

Salles, J.F., Pereira e Silva, M.C.P., Dini-Andreote, F., Dias, A.C., Guillaumaud, N., Poly, F., and Van Elsas, J.D. (2017). Successional patterns of key genes and processes involved in the microbial nitrogen cycle in a salt marsh chronosequence. Biogeochemistry *132*, 185–201.

Schimel, J.P., Gulledge, J.M., Clein-Curley, J.S., Lindstrom, J.E., and Braddock, J.F. (1999). Moisture effects on microbial activity and community structure in decomposing birch litter in the Alaskan taiga. Soil Biol. Biochem. *31*, 831–838.

Schipper, L.A., Degens, L.B., Sparling, G., and Duncan, L. . (2001). Changes in microbial heterotrophic diversity along five plant successional sequences. Soil Biol. Biochem. 33, 2093–2103.

Schmidt, S.K., Reed, S.C., Nemergut, D.R., Grandy, A.S., Cleveland, C.C., Weintraub, M.N., Hill, A.W., Costello, E.K., Meyer, A.F., Neff, J.C., *et al.* (2008). The earliest stages of ecosystem succession in high-elevation (5000 metres above sea level), recently deglaciated soils. Proc. Biol. Sci. *275*, 2793–2802. https://doi.org/10.1098/rspb.2008.0808

Schmidt, S.K., Nemergut, D.R., Darcy, J.L., and Lynch, R. (2014). Do bacterial and fungal communities assemble differently during primary succession? Mol. Ecol. *23*, 254–258.

Schrama, M., Berg, M.P., and Olff, H. (2012). Ecosystem assembly rules: the interplay of green and brown webs during salt marsh succession. Ecology 93, 2353–2364.

Schrama, M., Jouta, J., Berg, M.P., and Olff, H. (2013). Food web assembly at the landscape scale: using stable isotopes to reveal changes in trophic structure during succession. Ecosystems *16*, 627–638.

Shade, A., Caporaso, J.G., Handelsman, J., Knight, R., and Fierer, N. (2013). A meta-analysis of changes in bacterial and archaeal communities with time. ISME J. *7*, 1493–1506. https://doi.org/10.1038/ismej.2013.54

Shi, Y., Tyson, G.W., and DeLong, E.F. (2009). Metatranscriptomics reveals unique microbial small RNAs in the ocean's water column. Nature *459*, 266–269. https://doi.org/10.1038/nature08055

Sigler, W.V., and Zeyer, J. (2002). Microbial diversity and activity along the forefields of two receding glaciers. Microb. Ecol. *43*, 397–407. https://doi.org/10.1007/s00248-001-0045-5

Sigler, W.V., Crivii, S., and Zeyer, J. (2002). Bacterial succession in glacial forefield soils characterized by community structure, activity and opportunistic growth dynamics. Microb. Ecol. *44*, 306–316. https://doi.org/10.1007/s00248-002-2025-9

Staley, J.T., and Konopka, A. (1985). Measurement of in situ activities of nonphotosynthetic microorganisms in aquatic and terrestrial habitats. Annu. Rev. Microbiol. *39*, 321–346. https://doi.org/10.1146/annurev.mi.39.100185.001541

Stevenson, B.S., and Schmidt, T.M. (2004). Life history implications of rRNA gene copy number in Escherichia coli. Appl. Environ. Microbiol. *70*, 6670–6677.

Sutherland, W.J., Freckleton, R.P., Godfray, H.C.J., Beissinger, S.R., Benton, T., Cameron, D.D., Carmel, Y., Coomes, D.A., Coulson, T., Emmerson, M.C. *et al.* (2013). Identification of 100 fundamental ecological questions. J. Ecol. *101*, 58–67.

Torsvik, V., Øvreås, L., and Thingstad, T.F. (2002). Prokaryotic diversity – magnitude, dynamics, and controlling factors. Science *296*, 1064–1066. https://doi.org/10.1126/science.1071698

Tringe, S.G., von Mering, C., Kobayashi, A., Salamov, A.A., Chen, K., Chang, H.W., Podar, M., Short, J.M., Mathur, E.J., Detter, J.C., *et al.* (2005). Comparative metagenomics of microbial communities. Science *308*, 554–557.

Tscherko, D., Rustemeier, J., Richter, A., Wanek, W., and Kandeler, E. (2003). Functional diversity of the soil microflora in primary succession across two glacier forelands in the Central Alps. Eur. J. Soil Sci. *54*, 685–696.

Van Elsas, J.D., Trevors, J.T., Jansson, J.K., and Nannipieri, P. (2019). Modern Soil Microbiology, 3rd edn. (Taylor & Francis, New York).

Venter, J.C., Remington, K., Heidelberg, J.F., Halpern, A.L., Rusch, D., Eisen, J.A., Wu, D., Paulsen, I., Nelson, K.E., Nelson, W., *et al.* (2004). Environmental genome shotgun sequencing of the Sargasso Sea. Science *304*, 66–74. https://doi.org/10.1126/science.1093857

Verberkmoes, N.C., Russell, A.L., Shah, M., Godzik, A., Rosenquist, M., Halfvarson, J., Lefsrud, M.G., Apajalahti, J., Tysk, C., Hettich, R.L., *et al.* (2009). Shotgun metaproteomics of the human distal gut microbiota. ISME J. *3*, 179–189. https://doi.org/10.1038/ismej.2008.108

Vogel, T.M., Simonet, P., Jansson, J.K., Hirsch, P.R., Tiedje, J.M., Van Elsas, J.D., Bailey, M.J., Nalin, R., and Philippot, L. (2009). TerraGenome: a consortium for the sequencing of a soil metagenome. Nat. Rev. Microbiol. 7, 252.

Vos, M., Wolf, A.B., Jennings, S.J., and Kowalchuk, G.A. (2013). Micro-scale determinants of bacterial diversity in soil. FEMS Microbiol. Rev. 37, 936–954. https://doi.org/10.1111/1574-6976.12023

Walker, L.R., and Del Moral, R. (2003). Primary succession and ecosystem rehabilitation (Cambridge University Press, Cambridge).

Walker, L.R., Wardle, D.A., Bardgett, R.D., and Clarkson, B.D. (2010). The use of chronosequences in studies of ecological succession and soil development. J. Ecol. *98*, 725–736.

Wang, M., Yang, P., and Falcão Salles, J. (2015). Distribution of root-associated bacterial communities along a salt-marsh primary succession. Front. Plant Sci. *6*, 1188. https://doi.org/10.3389/fpls.2015.01188

Weber, C.F., and King, G.M. (2010). Distribution and diversity of carbon monoxide-oxidizing bacteria and bulk bacterial communities across a succession gradient on a Hawaiian volcanic deposit. Environ. Microbiol. *12*, 1855–1867. https://doi.org/10.1111/j.1462-2920.2010.02190.x

Whitman, W.B., Coleman, D.C., and Wiebe, W.J. (1998). Prokaryotes: the unseen majority. Proc. Natl. Acad. Sci. U.S.A. *95*, 6578–6583.

Yano, K., Wada, T., Suzuki, S., Tagami, K., Matsumoto, T., Shiwa, Y., Ishige, T., Kawaguchi, Y., Masuda, K., Akanuma, G., et al. (2013). Multiple rRNA operons are essential for efficient cell growth and sporulation as well as outgrowth in Bacillus subtilis. Microbiology *159*, 2225–2236.

Yeoh, Y.K., Dennis, P.G., Paungfoo-Lonhienne, C., Weber, L., Brackin, R., Ragan, M.A., Schmidt, S., and Hugenholtz, P. (2017). Evolutionary conservation of a core root microbiome across plant phyla along a tropical soil chronosequence. Nat. Commun. *8*, 215. https://doi.org/10.1038/s41467-017-00262-8

Young, I.M., Crawford, J.W., Nunan, N., Otten, W., and Spiers, A. (2008). Microbial distribution in soils: physics and scaling. Adv. Agron. *100*, 81–121.

Zhou, Z., Wang, C., Jiang, L., and Luo, Y. (2017). Trends in soil microbial communities during secondary succession. Soil Biol. Biochem. *115*, 92–99.

Insular Microbiogeography: Three Pathogens as Exemplars

4

James H. Kaufman[1*], Christopher A. Elkins[2,3], Matthew Davis[1], Allison M. Weis[4], Bihua C. Huang[4], Mark K. Mammel[2], Isha R. Patel[2], Kristen L. Beck[1], Stefan Edlund[1], David Chambliss[1], Judith Douglas[1], Simone Bianco[1], Mark Kunitomi[1] and Bart C. Weimer[4*]

[1]IBM Almaden Research Center, San Jose, CA, USA.
[2]Division of Molecular Biology, Center for Food Safety and Applied Nutrition, United States Food and Drug Administration, Laurel, MD, USA.
[3]Division of Healthcare Quality Promotion, National Center for Emerging and Zoonotic Infectious Diseases, CDC, Atlanta, GA, USA.
[4]University of California Davis, School of Veterinary Medicine, 100K Pathogen Genome Project, Davis, CA, USA.

*Correspondence: jhkauf@us.ibm.com; bcweimer@ucdavis.edu

https://doi.org/10.21775/9781912530021.04

Abstract

Traditional taxonomy in biology assumes that life is organized in a simple tree. Attempts to classify microorganisms in this way in the genomics era led microbiologists to look for finite sets of 'core' genes that uniquely group taxa as clades in the tree. However, the diversity revealed by large-scale whole genome sequencing is calling into question the long-held model of a hierarchical tree of life, which leads to questioning of the definition of a species. Large-scale studies of microbial genome diversity reveal that the cumulative number of *new* genes discovered increases with the number of genomes studied as a power law and subsequently leads to the lack of evidence for a unique core genome within closely related organisms. Sampling 'enough' new genomes leads to the discovery of a replacement or alternative to any gene. This power law behaviour points to an underlying self-organizing critical process that may be guided by mutation and niche selection. Microbes in any particular niche exist within a local web of organism interdependence known as the microbiome. The same mechanism that underpins the macro-ecological scaling first observed by MacArthur and Wilson also applies to microbial communities. Recent metagenomic studies of a food microbiome demonstrate the diverse distribution of community members, but also genotypes for a single species within a more complex community. Collectively, these results suggest that traditional taxonomic classification of bacteria could be replaced with a quasispecies model. This model is commonly accepted in virology and better describes the diversity and dynamic exchange of genes that also hold true for bacteria. This model will enable microbiologists to conduct population-scale studies to describe microbial behaviour, as opposed to a single isolate as a representative.

Introduction

For over 280 years biologists have classified living organisms using a system first proposed by Carl Linnaeus. This system developed as scientists grouped organisms based on observable characteristics (phenotypes), usually biochemical reactions, into a hierarchical taxonomic tree. This model has guided research ever since (Ruggiero *et al.*, 2015). Today, for the first time, microbiologists are engaged in describing the diversity of microbial life on earth utilizing technological advances in high-throughput sequencing and whole genome sequences (WGS). To obtain a bacterial genome, microbiologists use pure culture techniques to grow and isolate colonies that were commonly called 'clonal', prior to WGS, because the handful of selected biochemical tests were identical among the isolates. Extracting DNA from these colonies, high-throughput sequencing is used to obtain a consensus or average genome (Draper *et al.*, 2017) that is often described to be the prototypical genome. With a conventional world view of taxonomy, this 'clonal' genome is considered to define the organism's taxonomic class (i.e. species) and is thought to have a core or conserved set of genes that can be used to ascribe genes (i.e. traits) for use as unique identifiers that provide a basis for creating a group or clade on a hierarchal tree.

As WGS provides more data, it has become clear that isolated colonies are not clonal, and outbreaks are not represented by clonal organisms; rather, outbreaks are associated with a distribution of pathogenic genotypes. The observed diversity calls into question the concept of a species if enough diversity is captured from various sources of that specific organism. It is possible that the genome diversity is so high that the concept of a prototypical genome that represents a species or even serotype cannot be associated with a single conserved genome. As ever more bacterial genomes are sequenced, more variants are discovered. The taxonomy designed to classify them continues to fragment into ever more classes if the current hierarchal tree approach is maintained.

This diversity has driven the use of WGS to produce reference sequences for use in outbreak identification and investigation. A benefit to this work is production of vast numbers of data accessible for scientific research and comparison that are now uncovering evidence of genome diversity on a scale we have not contemplated. There has been an unprecedented increase in WGS and public release of new genomes is approaching 100,000 per year – a scale that was not anticipated 5 years ago. The repositories are often public and include the National Center for Biotechnology Information (NCBI), Sequence Read Archive (SRA), European Nucleotide Archive (ENA), DNA Data Bank of Japan (DDBJ), Genomic Encyclopedia of Bacteria and Archaea (GEBA), and multinational 100K Pathogen Genome Project (Allard, 2015; Weimer, 2012, 2107; Wu *et al.*, 2009). The crowd-sourced data in these WGS databases has resulted in cataloging of a plethora of microbial genomes and has assisted in the examination of the metagenomics of ecological niches important in medicine, agriculture, energy, and the built environment (Alivisatos *et al.*, 2015; Locey and Lennon, 2016; Shapiro *et al.*, 2012), all of which relies on reference genome databases.

The implementation of large-scale genomics in microbiology gives rise to observations previously impossible from the observations of a few isolates. In particular, the observed genome diversity challenges long-held beliefs about genome stability, evolutionary rate, and even the definition of a species (Doolittle, 1999; Doolittle and Brunet, 2016). Indeed, the number of unique genes that are represented within a species, known as the 'pan-genome', continues to increase as the expected number of genes conserved across all the members of a species, known as the 'core-genome', diminishes. This powers a new ability to define the trajectory of novel gene discovery to evoke consideration of the advances made by naturalists and applied to animals and theories of animal ecology. With the scale of genomes available the same considerations can be made when examining the diversity of microbial life and microbiome membership. Application of ecological theory provides a new perspective on microbiome structure and enables new insights into community association and membership that have not been examined in detail.

On a macroscopic scale, MacArthur and Wilson's macro-ecological 'theory of island biogeography' is a well-established explanation for species diversity across geographically disconnected groups (Diamond, 1984; Lovejoy *et al.*, 1984, 1986; MacArthur and Wilson, 1963). This theory relates the number of species (S) found, in steady state, within the area (A) of an isolated ecosystem as a power law. Over five orders of area magnitude, S varies as A^z with z, the MacArthur–Wilson exponent, typically in the range of $0.2 < z < 0.3$, decreasing slightly for island groups nearer to continental land masses that enable a slow exchange of species.

MacArthur-Wilson attributes the steady state species–area relationship to the rate of successful immigration of new species (from distant continental reservoirs) and the rate of species extinction, both dependent on the availability of viable niches (MacArthur and Wilson, 1963). The number of niches depends, in turn, on the available land area. Wilson's work also demonstrates that the surviving organisms can, themselves, provide niches for other organisms in the community, including new migrants (Wilson, 1999). As such, the observed scale-free behaviour in macro-ecology depends not only on the geographical distribution of land (Mandelbrot, 1983), but also on the dynamic evolution of the food web (Wilson, 1999). Indeed, ecological models and simulations based on Wilson's work reveal that scale-free behaviour emerges *only* as the food web develops over time (Bak *et al.*, 1988; Kaufman *et al.*, 1998). This process is precisely the mechanism that Bak, Tang, and Wiesenfeld (BTW) predicted would lead to power law scaling in their theory of 'self-organized criticality' (Bak *et al.*, 1988; Mandelbrot, 1983).

Self-organized criticality (SOC) is a ubiquitous mechanism by which complexity arises in nature. The behaviours predicted by the SOC theory have been observed across a wide variety of fields including physics, geology, ecology, and neuroscience (Kaufman *et al.*, 1998; Linkenkaer-Hansen *et al.*, 2001; Smalley *et al.*, 1985). Bak *et al.* (1988) propose that self-organized criticality is a property of any dynamical system that has a critical point *as an attractor*. Such a critical point is often observable experimentally as power law divergence of a correlation length of the system. In physics, critical behaviour often indicates a second-order phase transition. For example, the susceptibility of a ferromagnet diverges as the correlation length of magnetic moments diverges near the critical 'Curie' temperature (Kittel and Holcomb, 1967). Near a critical point, temporal and spatial system variables exhibit power law correlations. The value of the critical exponents for different variables are not the same, but they are not independent either. In fact they are related by simple scaling relations (Tang and Bak, 1988). Dynamical systems exhibit SOC if scaling emerges without the need to tune a control variable or effective temperature. This is the basis for the name of the theory. The system is self-organized because it tunes itself. The critical point is an attractor of the dynamics. Typically, SOC systems are non-equilibrium but slowly driven (Bak and Paczuski, 1995). Evolutionary systems with low mutation rates satisfy this criterion. Typically, SOC systems have many degrees of freedom such as an ecosystem with a complex web of interdependence.

Metagenomic analysis and microbial life: a large-scale study

In this chapter we describe a large-scale study of gene diversity for thousands of public genomes and three different bacterial genera. Completed in collaboration with our researchers at IBM, the University of California Davis, and the US Food and Drug Administration, this work consists of experiments that quantify the diversity of genes, confirm the power law relationship, redefine taxonomic context, and invoke the quasispecies model for bacteria. To test these principles at the scale of genes and genomes, one must have access to an enormous number of individuals (e.g. WGS) that reflect species diversity. The public databases of WGS provide the underlying data diversity. We took advantage of this large collection to examine the diversity for thousands of public genomes and three different genera to test the applications of BCW theory in the microbial world.

A growing library of genetic data

All WGS data are publicly available via NCBI (Weimer, 2012). The WGS were downloaded, assembled using ABySS (abyss-pe v.1.5.2) (Simpson *et al.*, 2009), and annotated using Prokka (v.1.10) (Seemann, 2014) prior to the analysis. These include new WGS data, sequenced by the 100K Pathogen Genome Project (UC Davis, Davis, CA; Weimer, 2107), as described by Lüdeke *et al.* (2015), using Illumina paired end 100 methods, publicly released on the SRA in the 100K Pathogen Genome Project bioproject (PRJNA186441) and re-assembled for use in this study. To measure the cumulative rate of observation of 'new' genes (both known and 'putative' or unknown) as a function of number of *Salmonella* isolates sampled, 866 individual *Salmonella* isolates were obtained from the 100K Pathogen Genome Project with *de novo* assembly with random subsets of those isolates selected in separate trials, followed by cumulative gene count determined for each trial.

To measure the cumulative rate of observation of 'new' *Campylobacter* genes as a function of the number of isolates sampled, the raw sequence data from 15,158 individual *Campylobacter* isolates from public sources (Leinonen *et al.*, 2010) were assembled and annotated as described for *Salmonella*. The number of genomes represents all available *Campylobacter* WGS data in the SRA, at the time of this publication, for which assembly was successful. The large number was chosen to test the observed scale-free behaviour over four orders of magnitude in the bootstrap as a function of the number of genomes.

Identification of core genes to build the phylogeny for *E. coli* was carried out with the Basic Local Alignment Search Tool (BLAST) with a criterion of 95% identity over 90% of the gene length, querying 42 closed reference genomes. An allowance was made whereby a core gene was retained if missing in no more than one genome. Alignment of the 3348 strain sequences for each gene yielded a single nucleotide polymorphism (SNP) set that was used to construct a phylogeny. From the phylogeny, 334 genotypes were selected to represent the diversity in the tree.

Whole RNA metagenomes were produced using HiSeq 4000 or X instruments with one sample per lane, resulting in ~350 million reads/sample. Genomic distances were determined using the Meier-Kolthoff method (Meier-Kolthoff *et al.*, 2013). Whole genome–genome distance matrices were translated into the Newick tree format using Mega7 with the neighbour-joining method. Genome distance matrixes were clustered and visualized using Matlab and R (Kumar *et al.*, 2016). Genome–genome similarity is defined as 1.0 – genome–genome distance.

Diversity of genes

To quantify genome diversity, we examined the specific gene content by comparing individual proteins from thousands of genomes from three different genera that occupy similar niches in animals: *Salmonella, E. coli,* and *Campylobacter.* All genomes were derived from publicly available WGS datasets where each organism was cultured, isolated, and sequenced. We performed a *de novo* assembly using ABySS (abyss-pe v.1.5.2) (Simpson *et al.*, 2009) and gene annotation using Prokka (v.1.10) (Seemann, 2014) for each genome and then quantified the cumulative number of total genes (i.e. proteins) discovered as a function of number of isolates (genomes) sampled in multiple bootstrapped trials (Fig. 4.1).

Each point in Fig. 4.1 represents the cumulative number of distinct amino acid sequences (proteins) inferred for regions annotated as known or hypothetical proteins for one genome, plotted against the number of genomes or isolates in that subset, both known and unknown. The points were derived from 100 separate bootstrapped trials.

For the 866 individual *Salmonella* genomes obtained and re-assembled, the log of the genome diversity grows approximately linearly with the log of the number of isolates (Fig. 4.1a), indicating a power law relationship with the average rate of cumulative gene discovery, *N*, increasing with the number of isolates, η, as shown in Equation 4.1:

$$N \propto \eta^{0.468 \pm 0.001} \; (Salmonella) \quad (4.1)$$

The same power law (within experimental error) was observed in an analysis of *E. coli* genomes (Fig. 4.1b). In this experiment, 3348 independent genome sequences were selected by classifying them into 334 genotypes to condense the data and search for a common reference representative with < 0.2% genetic difference. As with *Salmonella* (Fig. 4.1a), a power law emerged for *E. coli* relating the average rate of cumulative gene discovery, *N*, to the number of genotypes, η, as shown in Equation 4.2:

$$N \propto \eta^{0.462 \pm 0.002} \; (E.\ coli) \quad (4.2)$$

To determine whether genome size and/or mutation rate contributes to variation in the exponent, we extended this investigation to another organism, *Campylobacter* (Fig. 4.1c). *Campylobacter* is often found in the same niche (e.g. chicken microbiome) and is known to have a much higher mutation rate than either *E. coli* or *Salmonella* (which have similar mutation rates). To compensate for the additional genetic inclusion rates, we increased the number of genomes to 15,158. This greatly expanded the scale of the genomic calculation yet produced a similar exponent to the other organisms, again demonstrating that genome diversity followed an approximate power law relating the rate of cumulative gene discovery, *N*, to the number of genotypes, η, as shown in Equation 4.3:

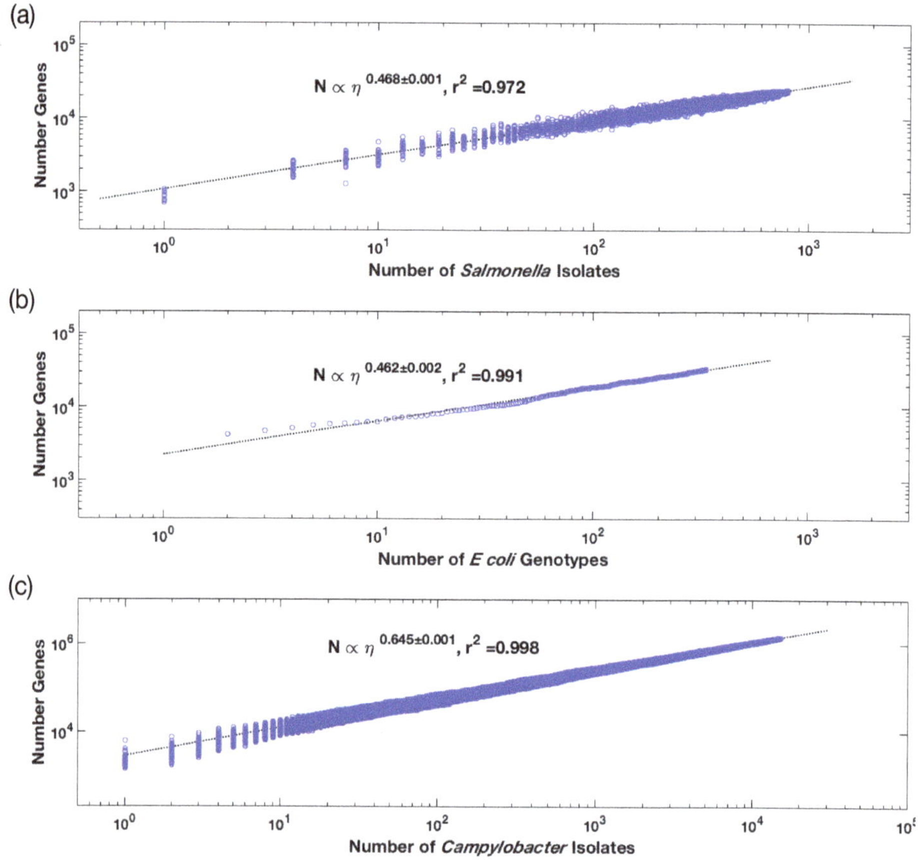

Figure 4.1 The cumulative rate of discovery of 'new genes' as a function of the number of (a) 866 *Salmonella* isolates, (b) *E coli* genotypes from 3348 representing 334 genotypes, and (c) 15,158 *Campylobacter* isolates. All increase as a power law (linear on a log-log plot) defined by Equations 4.1–4.3.

$$N \propto \eta^{0.645 \pm 0.001} \ (Campylobacter) \tag{4.3}$$

In Fig. 4.1, we deliberately use the same symbol (η) to denote both number of isolates (Fig. 4.1a andc) and number of genotypes (Fig. 4.1b). Each genotype represents a collection of isolates grouped in a data reduction effort. The power law behaviour observed in these studies of three different genera at completely different scales suggests a possibly universal functional relationship between the rate of new gene discovery and the number of genotypes or isolates over a wide range in scale. This type of scale-invariant behaviour does not spontaneously emerge in nature, but must reflect an underlying dynamical (evolutionary) process (Bak *et al.*, 1988).

Taxonomic content

As the number of bacterial WGS increases to over 330,000 in public databases, the emerging genotypic diversity is challenging traditional taxonomic hierarchical structures and concepts of evolution (Al-Saari *et al.*, 2015). Classification and naming of organisms with traditional methods is becoming problematic, as the addition of a single new genome drives continual revision of reference phylogenetic trees. For example, recently discovered cryptic environmental lineages of *E. coli* do not fit with current multi-locus sequence typing (MLST) classification for *E. coli sensu stricto* (Walk *et al.*, 2009). WGS analysis with a single gene, such as 16s rRNA, provides one name, but whole genome alignment provides a mixture of names where portions of the genome align to more than one genome, demonstrating genome plasticity, widespread homology among gene groups, and possible horizontal gene transfer at a much higher frequency that previously observed.

A similar conclusion was hinted at in a comparative study of lactic acid bacteria that led to the re-alignment of two entire genera (Makarova *et al.*, 2006). As the number of WGS data continues to grow at an unprecedented rate, conventional methods to classify organisms using genomics may not meet the standard of absolute identification based on phenotype. Whole genome comparison methods now available circumvent the assumptions currently used to define a constant or static set of single genes, SNPs, or even core sets of genes. Whole genome comparison eliminates the need for data reduction and avoids the associated skewing of taxonomical classification. Although there is a regulatory need to classify and name organisms, the data suggest that classification of bacteria, based on whole genomes, should be considered more akin to the classification of virus in terms of quasispecies, as discussed below. Methods that identify new organisms from metagenome sequence are emerging, making it impossible to use phenotypic characteristics as there is no pure culture (Parks *et al.*, 2017). These advances are begging for new methods to classify organisms based on genotype.

To illustrate how whole genome data challenge traditional taxonomy, it is informative to carry out whole genome comparison at different scales. Fig. 4.2 is an analysis of three independent *Campylobacter* datasets.

Principal component analysis (Jolliffe, 2002) based on a whole genome–genome difference matrix reveals similar clusters of *Campylobacter* genomes (Weimer, 2012; Weis, 2016). When using only 90 genomes from a study of primates and crows (Fig. 4.2a), three tight clusters appear in the PCA biplot; this result aligns well with classic phylogenetic classification and nomenclature and is congruent with a small sampling of the genome space based on only a few genomes. Using 218 genomes from the Ensembl database (Fig. 4.2b), the clusters for the three species in Fig. 4.2a enlarge and overlap substantially, demonstrating the increased genome diversity for an increase of approximately two times more genomes in the analysis. With 715 genomes from the NCBI SRA (Fig. 4.2c) with the same principal component analysis, the *Campylobacter* genome diversity becomes even more apparent. Taken together, these data demonstrate a rapid expansion of genotypic diversity within species, and of diversification, that overlaps species in same genome space. The whole genome analysis reveals how a relatively small increase in the number of genomes can rapidly expand genome space and blur the concept of species as well as traditional classification structures.

This expansion of genome space reflects the increasing genome diversity that occurs as more genomes are added to an analysis. Expansion of the diversity is especially evident when samples originate from a large collection of organisms from different geographical locations that may be insular regions of diversity (MacArthur and Wilson, 1963; Weis, 2016). The expansion and overlap in Fig. 4.2 also implies that traditional naming conventions are inaccurate and lead to false identification based on artefacts that emerge when using a small set

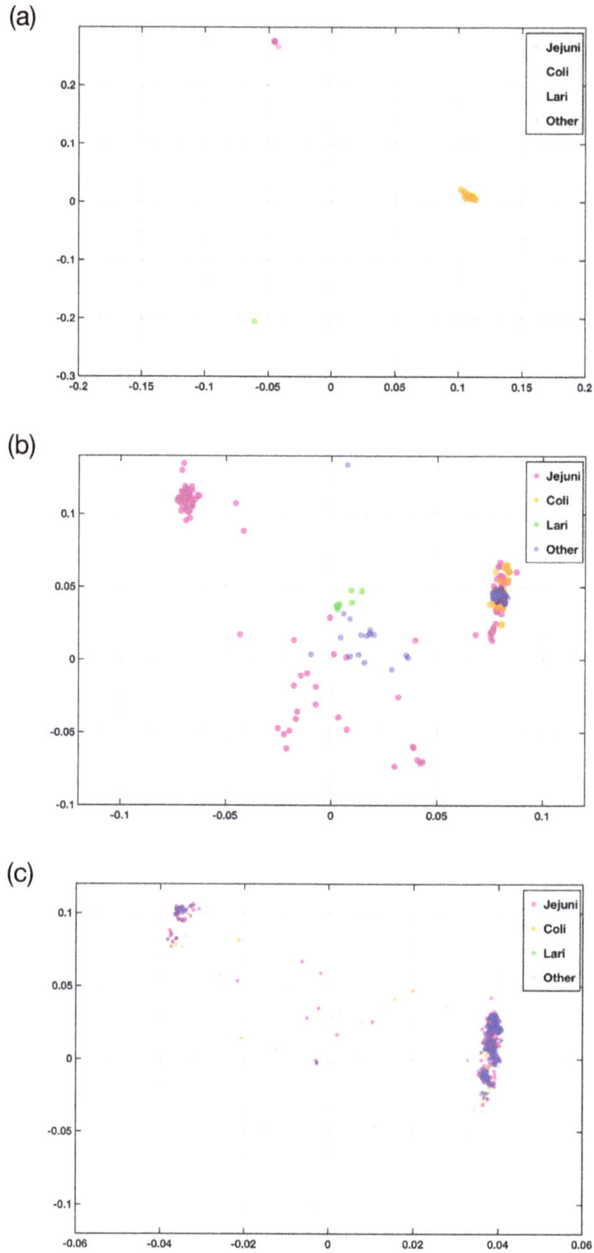

Figure 4.2 Principal component analyses of genome–genome distance from *Campylobacter* genomes representing (a) 90 genomes from primates and crows (Weis, 2016), (b) 218 genomes from the Ensembl reference database, and (c) 715 genomes from the NCBI SRA. In all three figures the *x*-axis is principal component 1 and the *y*-axis is principal component 2. The colour code in the legend indicates the original source species identification. With increasing dataset size, the clades representing traditional views of phylogeny for species begin to expand and overlap.

of input genes, MLST markers, or SNPs. These approaches were useful when computation and data were a limiting factor; however, with more genomes and cloud computing, data reduction methods are not needed and lead to inaccurate or even misleading classification compared with the use of whole genomes and whole genome distance.

Another way to visualize taxonomic grouping is to render a heat map of the full genome–genome distance matrix. The head map is shown in Fig. 4.3, which represents all pairwise distances between *Campylobacter* whole genomes.

With 90 genomes (Fig. 4.3a) the regions of greatest similarity are found along the matrix diagonal. With 218 and 715 genomes (Fig. 4.3b and c, respectively), highly similar but divergent subsets of genomes emerge with increasing numbers of genomes. These appear

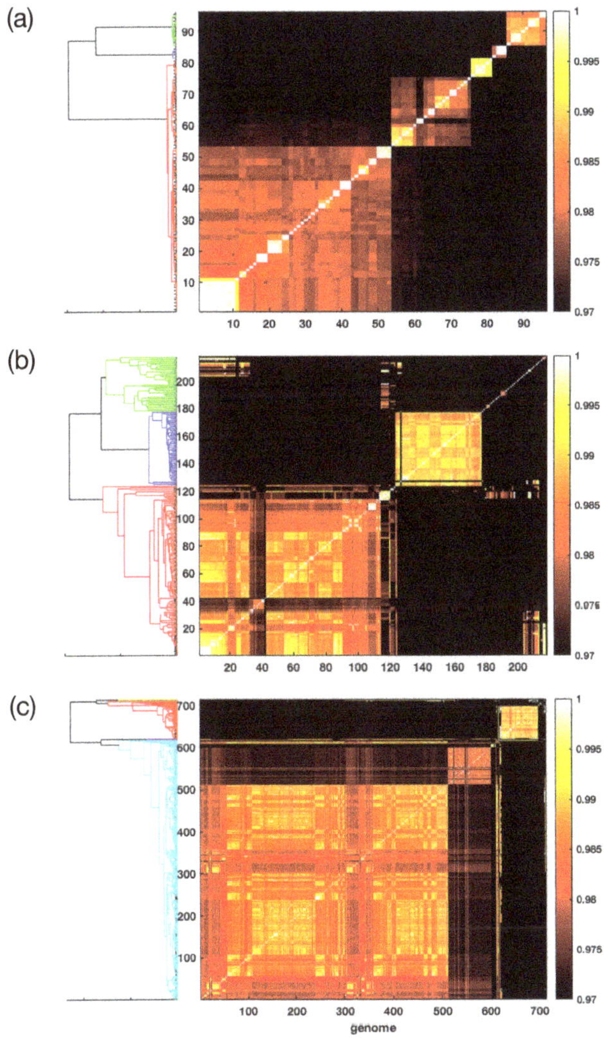

Figure 4.3 Heat maps showing the genome–genome similarity matrix derived from three *Campylobacter* databases containing (a) 90 genomes, (b) 218 genomes, and (c) 715 genomes.

as regions of similarity for well-separated (off-diagonal) genome pairs, indicating similarity between divergent branches of the taxonomic graph. This off-axis similarity reflects processes such as horizontal gene transfer that can result in distantly related organisms sharing similar or identical genes.

When 715 genomes were compared (Fig. 4.3c), the matrix reveals regions of genetic similarity for even more distant taxa, indicating gene transfer between different species.

Analogous behaviour is observed in divergent branches of *E. coli* and *Salmonella enterica* as evidenced by FDA microarray data, first reported by a number of other studies (Elkins *et al.*, 2013; Jackson *et al.*, 2011; Patel *et al.*, 2016).

Whole genome microarray hybridization using 1094 *E. coli* genomes from the FDA *E. coli* Identification (ECID) microarray (Fig. 4.4a) (Patel *et al.*, 2016) and 600 *S. enterica* profiled on a *S. enterica–E. coli* (SEEC) multispecies microarray (Fig. 4.4b) (Elkins *et al.*, 2013) reveal similar findings to the *in silico* analysis.

The majority of *E. coli* and *Salmonella* profiled by this method were amalgamated from clinical, foodborne, and associated environmental strains with relevance to public health and food safety with strains from private, academic, and publicly available collections. The microarray profiling program was developed by the FDA for track-and-trace molecular epidemiology (Elkins *et al.*, 2013). It was born out of an unconventional application towards genomic profiling from its original utility as a gene expression tool. This is primarily because forensic strain-level attribution became possible from the significant intraspecies genome plasticities observed in the resulting profiles (Jackson *et al.*, 2011). In aggregate, whole genome analyses reveal that individual species and strains could differ on the order of megabases using the entire genome. This amount of genomic variation raises questions about how to implement genome-based methods hierarchical classification methods, and points to a quasispecies approach for microbial forensics.

Shapiro *et al.* (2012) attributed this diversity expansion to horizontal gene transfer whereas Doolittle (1999) prophetically described gene sharing beyond what was thought

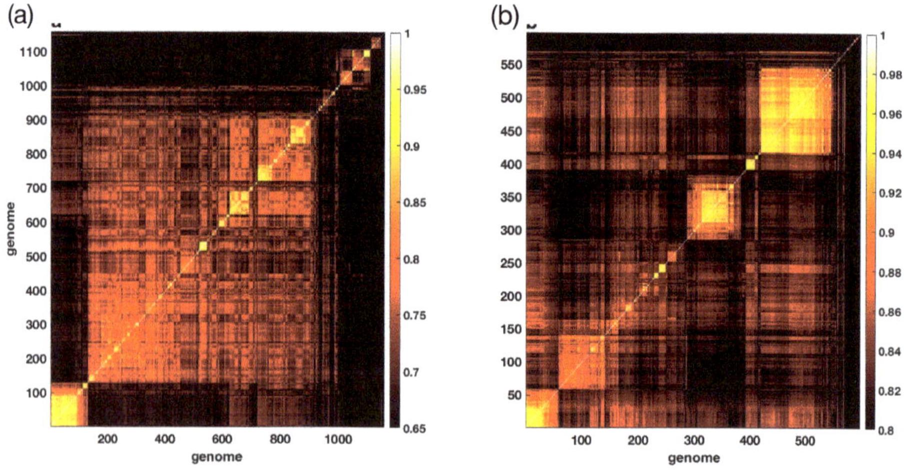

Figure 4.4 Heat maps of genome–genome similarity (Pearson correlations) of pangenome gene (allele) presence measured with custom-designed Affymetrix microarray platforms. (a) Hybridization intensities for 1094 *E. coli* genomes; (b) hybridization intensities for 600 *S. enterica* genomes.

possible. Subsequently, many studies have established that bacteria engage in extensive horizontal gene transfer (Doolittle, 1999; Doolittle and Brunet, 2016; Hug *et al.*, 2016; Woese, 2004). Horizontal gene transfer suggests that 'the tree of life' is not a simple tree (Hug *et al.*, 2016), but rather a complex web of genes that have uneven movement between organisms. The diversity apparent in Figs 4.1–4.4 suggests that bacterial life is organized in a highly interconnected network, a graph containing edges that connect phylogenetic quasispecies across varying genetic distances or scales.

The quasispecies model

Each of the thousands of public genomes analysed was extracted from some microbiome in which organisms existed in a web of organism interdependence. The communities were different, and the organisms and genes (or metagenomes) that existed depended on local conditions. In many cases, the local environment was defined by a macroscopic animal host. As such one might expect scaling behaviour observed in macro-ecologies to extend down to their microbiomes.

Our experiments reveal a power law scaling of the number of new genes versus the number of samples or genomes studied. Mathematically, this relationship rules out the existence of a core genome. As new genomes are studied, variations are eventually found for every gene. Some of these variants may represent slight modifications in genotype (single nonsynonymous mutations). Others represent entirely new phenotypes. We derive a scaling law below that relates the exponent measured here to the MacArthur and Wilson exponent, z. Our analyses of public genomes reveal a common exponent for closely related bacteria (*Salmonella* and *E. coli*) and a slightly larger exponent for a distant organism (*Campylobacter*) with a higher mutation rate and a smaller genome. All values of z are in agreement with the range observed by Wilson and MacArthur. This scaling behaviour reveals diversity that emerges from examining over 15,000 public samples. Experimental analysis of 40 metagenomic samples confirms large-scale genotypic diversity of *Campylobacter* in a single microbiome.

The scaling behaviour empirically observed requires us to rethink the system of classification used to define species. Eigen and Schuster suggest that a 'cloud' of diverse but related organisms within a population are more accurately represented as quasispecies with a distribution of genomes and a distribution of genes (Eigen and Schuster, 1977, 2012). This concept is commonly invoked in virology but data demonstrating its applications to bacterial evolution have been limited (Allard, 2015; Jolley *et al.*, 2004). Available genome sequences are just now reaching a critical mass required to make the observations demonstrated here (Allard, 2015; Leinonen *et al.*, 2010; Tateno *et al.*, 2002; Wu *et al.*, 2009).

The public genomes analysed here were submitted to the SRA by multiple researchers performing a wide variety of independent studies. In these studies, samples were derived using laboratory procedures where each pathogen was studied as a single genotype, reflecting a traditional view of phylogeny where organisms are cultured, isolated, and sequenced from single colonies. In fact, organisms exist in genomically diverse populations or communities of related, but not identical genotypes – even within a colony. These communities evolve rapidly under selective pressure by a number of mechanisms including horizontal gene transfer, methylation, and plasmid content. Culture organisms for sequencing as reference genomes are recognized as the *average genome of a colony*, in contrast to an oversimplified model of clonal outbreaks. In fact, these organisms exist in highly diverse populations of related genotypes that reproduce with high mutation rates including point mutations, larger scale insertions and deletions, and gene transfer. Shapiro *et al.* (2012) reported ecologically driven differentiation of genes in recently diverged populations of ocean bacteria. Although the number of samples in their study was not large enough to observe scale invariance, it demonstrates the rapidity with which gene transfer can occur across multiple related strains in response to environmental pressures: 'genomic fragments can sweep through populations in an ecology-specific manner ... with a clear bias towards within-habitat sharing of DNA' (Shapiro *et al.*, 2012).

The theory of quasispecies postulates an evolutionarily optimized mutation rate, which increases in response to stress. This provides for rapid adaptation including, for example, a very high rate of drug escape. Moreover, high recombination and horizontal gene transfer rates are known to play an important role in increasing both the adaptation rate and genetic diversity. Bacterial populations are known to exhibit high intrinsic mutation and recombination rates, throughout the course of an infection (Suerbaum and Josenhans, 2007) and/or during the acute phase of infection (Linz *et al.*, 2014). Increased mutation rates in bacteria are also apparent during selection pressure, either by natural predators (Weitz *et al.*, 2005) or through expression of alternative error-prone mutator genes (Ebrahimi-Rad *et al.*, 2003). Several *Campylobacter* strains are known to show increased mutation rates (Parkhill *et al.*, 2000), with instances of *C. jejuni* and *C. coli* hypermutator phenotypes linked to the emergence of ciproxin resistance.

The quasispecies model provides a framework to understand fitness and evolvability of a population (Jones *et al.*, 2015; Stern *et al.*, 2014; Xiao *et al.*, 2016). Fitness is determined not by the genetic characteristic of a single, isolated, static species or gene, but by the collective distribution of the members of the quasispecies whose clonal expansion reflects their underlying genotypic diversity and their fitness with respect to a particular environment or selection pressure including antibiotics, other medications, and treatments (Hu and Zhu, 2016; Qin *et al.*, 2010). The model that a particular pathogen associated with an epidemic or outbreak is accurately represented as a single species (e.g. a microbe's name using 16s sequencing) with a clonal identity is not only inadequate but also fallacious: it ignores the underlying genotypic diversity.

Metagenomic techniques make it possible to profile the genetic diversity of a quasispecies within a microbiome. We used RNAseq to study the metatranscriptome of the microbial community of 27 different poultry meal samples. Combining data from all samples provides ≈9.5B reads of raw sequence. Although metagenomics is often used to profile the community ecology, we used the data here to profile the genotypic diversity of one organism in the community, *Campylobacter*. These reads were aligned to the 218 *Campylobacter* genomes from the Ensembl database and the alignments tallied for each genome at 95% identity. This reference database is larger than is typically used to identity a single species in a metagenomic study, but small enough to illustrate two alternative views of what that identification means.

A classic analysis might treat the 218 *average genomes* in the reference database as independent genotypes, or clonal leaf nodes, on the tree of life. Using this model, if any of these genotypes were present in the community of the poultry meal, it would be possible to set a strict threshold and ignore any evidence of occurrence in the sample below that threshold.

An alternative analysis with a different perspective might treat the 218 genomes in the reference database as approximate, historic, average observations of genotypic *distributions* from 218 ecological niches. Alignments to this reference provide a measure of the probability with which genotypes in the *Campylobacter* quasispecies in the poultry meal are similar to those historic genotypes. Genotypes with several thousands of alignments may reflect minority alleles within the community.

Fig. 4.5 shows a circle plot mapping the log-normalized alignments from our metagenomic study of poultry meal to 218 *Campylobacter* reference genomes.

The log-normalization highlights the genotypic diversity reflected by both the reference database and the sample. In this view, it is still possible within the circle plot to label specific genotypes with alignment scores above a fixed threshold, but displaying the alignments to the full reference database reveals that there is, in fact, a distribution of closely related genotypes within the sample.

As WGS and metagenomics gain acceptance in fields like microbiome medicine, outbreak definition, and food safety traceability, where the demands for accuracy and precision are very high, what really matters is function because it is genes that drive the epidemiology. As shown in Fig. 4.1, hundreds of samples and tens of thousands of individual genes are required to reveal scaling behaviour and/or to estimate a power law exponent. To do this accurately requires inclusive and expansive reference databases. If too few references

Figure 4.5 A circle plot showing the mapping of all log-normalized read alignments from a metagenomic study of poultry meal to 218 *Campylobacter* genomes in the curated Ensembl database. Each circle represents one specific genome in the reference. The area of each circle is the log of the number of sequence reads that matched that genome. The arrangement of the small circles is arbitrary. The dominant *Campylobacter* species are indicated in the legend.

are in the database, inaccurate associations may be made leading to inappropriate attribution of cause and effect. These alignments reveal the genetic diversity of the *Campylobacter* quasispecies within the chicken meal microbiome based on the 218 curated genomes in Ensembl (Hubbard *et al.*, 2002).

Another approach to measuring diversity is to quantify the allelic variation observed in all reads that align to *Campylobacter*. To do this we enumerate all specific SNP variants and plot the frequency with which these occur (Fig. 4.6).

A particular *Campylobacter* SNP variant (where the nucleotide differs from the reference) that is found only once at a particular location is the most frequent event in the graph. Other variants, where the same substitution is found up to tens of thousands of times at a particular location, are found infrequently. The unique SNP variants were enumerated for all 218 genomes in the reference. The frequency of occurrence for the number of unique SNP variants is approximately a power law with slope near -1 (i.e. $1/f$).

We note that the evidence for scale invariance down to the level of single SNPs does not prove that all nucleotides within a genome are replaced with the same frequency or with the same scale-free behavior. Rather, the highest rate of substitution is typically observed near the 3′-ends of a gene. Substitutions near the 3′-ends are less likely to disrupt function, and more likely to survive (by natural selection).

Insular microbiogeography: an emerging concept for microbiology

The MacArthur–Wilson theory of island biogeography purports that there is an underlying power law relationship between species and habitat on a macroscopic level. Our experimental

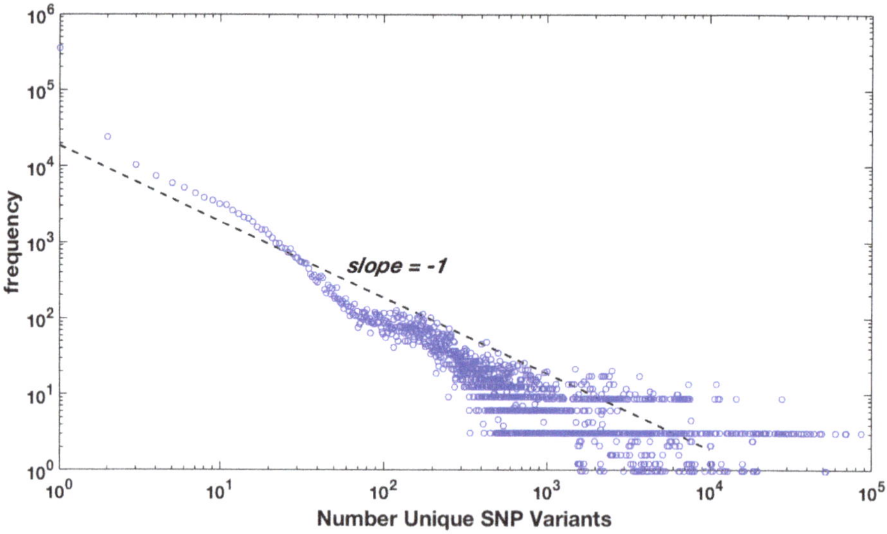

Figure 4.6 The occurrence frequency of unique *Campylobacter* genome SNP variants within a metaRNAseq study of a poultry meal. The function is approximately a power law with slope near -1.

results lead us to posit that the power law exponent that MacArthur and Wilson measured for macroscopic life also applies to microbial life. This is, by definition, scale-free behaviour and leads to some 'laws' that microbes live by to produce genomic variations that we can use as tools for many applications:

- Each host provides a unique environment for community of microorganisms. These microorganisms adapt to their environments.
- Because the distribution of host organisms obeys the MacArthur–Wilson law, so too will the genomic diversity of the microbial communities that colonize them.

This follows from the mathematical definition of scale-free behaviour and from ecological arguments. Microbial communities are linked by the larger web of hierarchical interdependence that links their macro-ecological hosts. To test this proposition, it remains to connect the exponent observed here to the MacArthur–Wilson exponent (Tang and Bak, 1988). The data above show the number of new genes compared with the number of samples or genotypes (Fig. 4.1), whereas MacArthur and Wilson's law relates species or taxa to *geographical area*. Moreover, the data here come primarily from sources originating from human patients and, if the human hosts are interpreted as a single niche, there may seem to be no reason to expect scaling behaviour or any diversity within the genotypic clades. Contrary to this expectation, we find the data indicate a possibly universal power law relationship as a function of number of genomes. In fact, medical sequencing of pathogens is indicated in the event of human disease, typically in unusual cases. As such, these reference sequences do not specifically represent *steady state populations*, but rather originate from consumption of contaminated food or water, or exposure to other disease vectors. Given a global food supply chain and global travel patterns, the simplest interpretation of the data is that these genomes represent a *near random sampling of foodborne pathogens* globally. Humans are engaged in this random sampling as part of everyday consumption. With random sampling, the number of isolates (or genotypes), η, in Fig. 4.3 is related to area, A, as a random walk through sample space. Formally, this implies a functional relationship between area and the number of genomes as $A \approx \eta^{1/2}$. The MacArthur–Wilson exponent, z, is then one-half the exponent obtained in Fig. 4.1 or:

$$N \propto \eta^{2z} \tag{4.4}$$

For all organisms studied here, the data put the corresponding MacArthur–Wilson exponent in the range $0.23 < z < 0.32$, well within the range observed for species-area scaling in the theory of island biogeography.

Equation 4.4 and the data in Fig. 4.1 indicate that as the number of genomes studied increases, the cumulative number of genes observed will continue to increase. The more genomes included in an analysis, the more genes will be discovered (Jacobsen *et al.*, 2011). Although our three experiments involved tens of thousands of genomes, none provided evidence for an upper limit to the rate of gene discovery. Such a limit would manifest as a 'knee' or bend on the log-log plots above some maximum of genomes. For example, in Fig. 4.1c, with 15,158 genomes and 1000 jackknife trials, the study spans a scale exceeding four orders of magnitude with no upper scaling limit found. These observations in three different organisms indicates that a core genome does not likely exist and that as more genomes are deposited the size of genome space will increase with no practical upper limit – therefore no boundary between species but rather a gradient of change from one to another.

The data in Fig. 4.1a–c reveal that the more genomes we sequence, the more genes will be discovered. For any taxonomic group, at the measured rate of gene discovery, the question becomes, 'How many samples would be required before an alternative gene is discovered for every gene in any randomly selected genome?'

The answer to this question is a special case of the 'coupon collector's problem' (Blom *et al.*, 2012). If we assume that non-synonymous mutations are possible for any gene, then the question becomes, after finding N new genes at random from a genome of length N_o, at what value of N does the probability of *not finding* an alternative gene for every gene in the original genome become vanishingly small.

Let $Z_i^N Z_i^r$ represent the event that the i-th gene is not replaced after N new genes are found. Then, i is r :

$$P\left(Z_i^N\right)=\left(1-\frac{1}{N_o}\right)^N \leq e^{-N/N_o} \tag{4.5}$$

where the inequality follows from the Taylor expansion of e^{-x} for small x.

For $N > 2\,N_o \ln(N_o)$:

$$P\left(N > 2\ N_o\ ln\left(N_o\right)\right)=P\left(\cup_i Z_i^{2\ N_o\ ln\left(N_o\right)}\right)\leq N_o P\left(Z_i^{2 N_o\ ln\left(N_o\right)}\right)\leq\frac{1}{N_o} \tag{4.6}$$

Therefore, the probability of not finding an alternative to *every gene* falls below $1/N_0$ for $N > 2N_0 \ln(N_0)$. For *Campylobacter*, with $N_0 \approx 1500$ genes, this requires finding $N \approx 22{,}000$ new genes. From Fig. 4.1 this will take place after randomly sampling just a few hundred isolates. For *Salmonella*, a variant to each of $N_0 \approx 4500$ genes will be found after discovering $\approx 76{,}000$ genes, which requires approximately 2000 isolates. However, this was not the case and the logical conclusion is that a unique core genome does not exist (DeLong *et al.*, 2010; Green *et al.*, 2004; Horner-Devine *et al.*, 2004).

Implications for microbiology

The evidence for scaling over several orders of magnitude in sample diversity of microbial genotypes and genomes has practical implications for microbiology and the way we catalogue organisms and investigate outbreaks. Approaches that depend on data reduction or definition of a core genome, selection of small numbers of SNPs, or of core MLST gene sets, are destined to fail with false negatives as they do not capture the expansive genome diversity that is unavoidable. The fact that adding one more genotype to the set can indicate a need to rebuild the taxonomy shows that we have not adequately sampled the genomic space and that the tree of life for bacteria is not a tree – it is a network (or graph) of interacting genomes within an environment or geography (Al-Saari *et al.*, 2015; Makarova *et al.*, 2006; Walk *et al.*, 2009), with loops that indicate gene transfer between divergent taxa.

As in virology, it is more appropriate to think of bacteria in terms of quasispecies with a distribution of genotypes and genes that evolve, move, and shift in genome space in response to a changing fitness landscape. The tools required for whole genome comparison using the entire genome exist and must be applied to capture genetic similarity between both nearby and distant genotypes. This is necessary for accurate identification and functional association required by applications in microbiome-induced physiology, outbreak detection, and food regulation. Adopting whole genome comparison may force microbiology to replace the practice of naming and even the concepts of 'genus' and 'species' with new data models and notations that label populations by their function and the niche(s) they occupy. From a regulatory and health perspective, it is most important to know if an organism has a pathogenic gene that may cause illness (regardless of its taxonomic classification).

There are several dimensions one might consider for a new nomenclature system. New notations could capture the diversity or statistical distributions that describe the frequency with which important function exists within a community. Community level techniques including metagenomics and meta-transcriptomics can be used to predict the probability that specific genes are (or are not) present in a community and the likelihood with which they may emerge as hazards in response to fitness pressures.

Conclusions: gene discovery and power law scaling

The discovery rate of new genes depends on number of genomes sequenced as a power law. If this scaling behaviour arises from the hierarchical web of species interdependence, one should expect the MacArthur–Wilson theory of insular biogeography to apply to microbes. MacArthur–Wilson never derived or explained the power law behaviour they discovered. That explanation comes from the more general theory of self-organized criticality (Bak *et al.*, 1988). The power law reflects the complexity of the underlying food web that connects the macroscopic and microscopic species with their chemical environment. Observation of the same scale-invariant behaviour for microbes and higher order life suggests this food web evolves towards a critical state with fluctuations expected on all scales. To test this proposition, we derive a scaling relation relating the observed exponent to the MacArthur–Wilson exponent, assuming that genome datasets in public repositories represent a near random sampling from the global genome space. The values obtained are consistent with theory. Species require niches to survive and thrive. With respect to microbes these niches are not restricted to the physical and chemical properties of insular environments (which themselves evolve a characteristic fractal geometry) but also include other species. It is the interacting web of ecological dependence between species that leads to self-organization, evolution of a scale-free diversity of life forms across scales within isolated ecosystems. The host organisms are linked to their microbiomes. They provide unique habitats for the communities of microbes that colonize them.

The diversity predicted by a theory of insular microbiogeography requires we rethink our approach to organism classification and regulation. Bacteria form quasispecies related by an slowly evolving phylogenetic graph – not a tree – that reflects the complex web of interdependence spanning all phylogenetic ranks down to the level of microbes. In any open ecosystem or open microbiome, the classical concept of core genomes ceases to be meaningful for any class of bacteria. Classification and regulation based on comparison of whole genomes and *gene function* may better capture the genotypic relationships between related microorganisms.

Acknowledgements

The authors would like to acknowledge contributions from and discussions with Nyuget Kong, Bob Baker, Peter Markwell, Dylan Storey, Barbara Jones and Kenneth L. Clarkson.

References

Al-Saari, N., Gao, F., Rohul, A.A., Sato, K., Sato, K., Mino, S., Suda, W., Oshima, K., Hattori, M., Ohkuma, M., *et al.* (2015). Advanced microbial taxonomy combined with genome-based-approaches reveals that *Vibrio astriarenae* sp. nov., an agarolytic marine bacterium, forms a new clade in Vibrionaceae. PLOS ONE *10*, e0136279. https://doi.org/10.1371/journal.pone.0136279

Alivisatos, A.P., Blaser, M.J., Brodie, E.L., Chun, M., Dangl, J.L., Donohue, T.J., Dorrestein, P.C., Gilbert, J.A., Green, J.L., Jansson, J.K., *et al.* (2015). MICROBIOME. A unified initiative to harness Earth's microbiomes. Science *350*, 507–508. https://doi.org/10.1126/science.aac8480

Allard, M. (2015). GenomeTrakr: A Pathogen Database to Build a Global Genomic Network for Pathogen Traceback and Outbreak Detection. Paper presented at 2015 Annual Meeting (July 25–28, 2015), DesMoines, IA, USA.

Bak, P., and Paczuski, M. (1995). Complexity, contingency, and criticality. Proc. Natl. Acad. Sci. U.S.A. *92*, 6689–6696.

Bak, P., Tang, C., and Wiesenfeld, K. (1988). Self-organized criticality. Phys. Rev. A *38*, 364.

Blom, G., Holst, L., and Sandell, D. (2012). Problems and Snapshots from the World of Probability (Springer Science & Business Media, Berlin).

DeLong, J.P., Okie, J.G., Moses, M.E., Sibly, R.M., and Brown, J.H. (2010). Shifts in metabolic scaling, production, and efficiency across major evolutionary transitions of life. Proc. Natl. Acad. Sci. U.S.A. *107*, 12941–12945. https://doi.org/10.1073/pnas.1007783107

Diamond, J. (1984). ' Normal' extinctions of isolated populations. In Extinctions, Nitecki, M.H., ed. (University of Chicago Press, Chicago), pp. 191–246.

Doolittle, W.F. (1999). Phylogenetic classification and the universal tree. Science *284*, 2124–2129.

Doolittle, W.F., and Brunet, T.D. (2016). what is the tree of life? PLOS Genet. *12*, e1005912. https://doi.org/10.1371/journal.pgen.1005912

Draper, J.L., Hansen, L.M., Bernick, D.L., Abedrabbo, S., Underwood, J.G., Kong, N., Huang, B.C., Weis, A.M., Weimer, B.C., van Vliet, A.H., *et al.* (2017). Fallacy of the unique genome: sequence diversity within single *Helicobacter pylori* strains. MBio *8*, e02321-16.

Ebrahimi-Rad, M., Bifani, P., Martin, C., Kremer, K., Samper, S., Rauzier, J., Kreiswirth, B., Blazquez, J., Jouan, M., van Soolingen, D., *et al.* (2003). Mutations in putative mutator genes of *Mycobacterium tuberculosis* strains of the W-Beijing family. Emerging Infect. Dis. *9*, 838–845.

Eigen, M., and Schuster, P. (1977). A principle of natural self-organization. Naturwissenschaften *64*, 541–565.

Eigen, M., and Schuster, P. (2012). The Hypercycle: A Principle of Natural Self-organization (Springer Science & Business Media, Berlin).

Elkins, C.A., Kotewicz, M.L., Jackson, S.A., Lacher, D.W., Abu-Ali, G.S., and Patel, I.R. (2013). Genomic paradigms for food-borne enteric pathogen analysis at the USFDA: case studies highlighting method utility, integration and resolution. Food Addit. Contam. Part A Chem. Anal. Control Expo. Risk Assess. *30*, 1422–1436. https://doi.org/10.1080/19440049.2012.743192

Green, J.L., Holmes, A.J., Westoby, M., Oliver, I., Briscoe, D., Dangerfield, M., Gillings, M., and Beattie, A.J. (2004). Spatial scaling of microbial eukaryote diversity. Nature *432*, 747–750.

Horner-Devine, M.C., Lage, M., Hughes, J.B., and Bohannan, B.J. (2004). A taxa-area relationship for bacteria. Nature *432*, 750–753.

Hu, Y., and Zhu, B. (2016). The human gut antibiotic resistome in the metagenomic era: progress and perspectives. Infect. Dis. Transl. Med. *2*, 41–47.

Hubbard, T., Barker, D., Birney, E., Cameron, G., Chen, Y., Clark, L., Cox, T., Cuff, J., Curwen, V., Down, T., *et al.* (2002). The Ensembl genome database project. Nucleic Acids Res. *30*, 38–41.

Hug, L.A., Baker, B.J., Anantharaman, K., Brown, C.T., Probst, A.J., Castelle, C.J., Butterfield, C.N., Hernsdorf, A.W., Amano, Y., Ise, K., *et al.* (2016). A new view of the tree of life. Nat. Microbiol. *1*, 16048. https://doi.org/10.1038/nmicrobiol.2016.48

Jackson, S.A., Patel, I.R., Barnaba, T., LeClerc, J.E., and Cebula, T.A. (2011). Investigating the global genomic diversity of *Escherichia coli* using a multi-genome DNA microarray platform with novel gene prediction strategies. BMC Genomics *12*, 349. https://doi.org/10.1186/1471-2164-12-349

Jacobsen, A., Hendriksen, R.S., Aaresturp, F.M., Ussery, D.W., and Friis, C. (2011). The *Salmonella enterica* pan-genome. Microb. Ecol. *62*, 487–504. https://doi.org/10.1007/s00248-011-9880-1

Jolley, K.A., Chan, M.S., and Maiden, M.C. (2004). mlstdbNet - distributed multi-locus sequence typing (MLST) databases. BMC Bioinformatics *5*, 86. https://doi.org/10.1186/1471-2105-5-86

Jolliffe, I. (2002). Principal component analysis (Wiley Online Library).

Jones, B.A., Lessler, J., Bianco, S., and Kaufman, J.H. (2015). Statistical mechanics and thermodynamics of viral evolution. PLOS ONE *10*, e0137482. https://doi.org/10.1371/journal.pone.0137482

Kaufman, J., Brodbeck, D., and Melroy, O. (1998). Critical biodiversity. Cons. Biol. *12*, 521–532.

Kittel, C., and Holcomb, D.F. (1967). Introduction to solid state physics. Am. J. Phys. *35*, 547–548.

Kumar, S., Stecher, G., and Tamura, K. (2016). MEGA7: Molecular Evolutionary Genetics Analysis version 7.0 for bigger datasets. Mol. Biol. Evol. *33*, 1870–1874. https://doi.org/10.1093/molbev/msw054

Leinonen, R., Sugawara, H., and Shumway, M. (2010). The sequence read archive. Nucleic Acids Res. *39*, D19–D21.

Linkenkaer-Hansen, K., Nikouline, V.V., Palva, J.M., and Ilmoniemi, R.J. (2001). Long-range temporal correlations and scaling behavior in human brain oscillations. J. Neurosci. *21*, 1370–1377.

Linz, B., Windsor, H.M., McGraw, J.J., Hansen, L.M., Gajewski, J.P., Tomsho, L.P., Hake, C.M., Solnick, J.V., Schuster, S.C., and Marshall, B.J. (2014). A mutation burst during the acute phase of *Helicobacter pylori* infection in humans and rhesus macaques. Nat. Commun. *5*, 4165. https://doi.org/10.1038/ncomms5165

Locey, K.J., and Lennon, J.T. (2016). Scaling laws predict global microbial diversity. Proc. Natl. Acad. Sci. U.S.A. *113*, 5970–5975. https://doi.org/10.1073/pnas.1521291113

Lovejoy, T.E., Rankin, J.M., Bierregaard, R., Brown, K.S., Emmons, L.H., and Van der Voort, M.E. (1984). Ecosystem decay of Amazon forest remnants. In Extinctions, Nitecki, M.H., ed. (University of Chicago Press, Chicago), pp. 295–325.

Lovejoy, T.E., Bierregaard, R.O., Rylands, A.B., Malcolm, J.R., Quintela, C.E., Harper, L.H., Brown, K.S., Powell, A.H., Powell, G.V.N., Schubart, H.O.R., *et al.* (1986). Edge and other effects of isolation on Amazon forest fragments. In Conservation Biology: The Science of Scarcity and Diversity, Soulé, M.E., ed. (Sinauer, Sunderland, MA), pp. 257–285.

Lüdeke, C.H., Kong, N., Weimer, B.C., Fischer, M., and Jones, J.L. (2015). Complete genome sequences of a clinical isolate and an environmental isolate of *Vibrio parahaemolyticus*. Genome Announc. *3*, e00216-15. https://doi.org/10.1128/genomeA.00216-15

MacArthur, R.H., and Wilson, E.O. (1963). An equilibrium theory of insular zoogeography. Evolution, 373–387.

Makarova, K., Slesarev, A., Wolf, Y., Sorokin, A., Mirkin, B., Koonin, E., Pavlov, A., Pavlova, N., Karamychev, V., Polouchine, N., *et al.* (2006). Comparative genomics of the lactic acid bacteria. Proc. Natl. Acad. Sci. U.S.A. *103*, 15611–15616.

Mandelbrot, B.B. (1983). The Fractal Geometry of Nature, Vol 173 (Macmillan, London).

Meier-Kolthoff, J.P., Auch, A.F., Klenk, H.P., and Göker, M. (2013). Genome sequence-based species delimitation with confidence intervals and improved distance functions. BMC Bioinformatics *14*, 60. https://doi.org/10.1186/1471-2105-14-60

Parkhill, J., Wren, B.W., Mungall, K., Ketley, J.M., Churcher, C., Basham, D., Chillingworth, T., Davies, R.M., Feltwell, T., Holroyd, S., *et al.* (2000). The genome sequence of the food-borne pathogen *Campylobacter jejuni* reveals hypervariable sequences. Nature *403*, 665–668. https://doi.org/10.1038/35001088

Parks, D.H., Rinke, C., Chuvochina, M., Chaumeil, P.A., Woodcroft, B.J., Evans, P.N., Hugenholtz, P., and Tyson, G.W. (2017). Recovery of nearly 8,000 metagenome-assembled genomes substantially expands the tree of life. Nat. Microbiol. *2*, 1533–1542. https://doi.org/10.1038/s41564-017-0012-7

Patel, I.R., Gangiredla, J., Lacher, D.W., Mammel, M.K., Jackson, S.A., Lampel, K.A., and Elkins, C.A. (2016). FDA *Escherichia coli* Identification (FDA-ECID) microarray: a pangenome molecular toolbox for serotyping, virulence profiling, molecular epidemiology, and phylogeny. Appl. Environ. Microbiol. *82*, 3384–3394. https://doi.org/10.1128/AEM.04077-15

Qin, J., Li, R., Raes, J., Arumugam, M., Burgdorf, K.S., Manichanh, C., Nielsen, T., Pons, N., Levenez, F., Yamada, T., *et al.* (2010). A human gut microbial gene catalogue established by metagenomic sequencing. Nature *464*, 59–65. https://doi.org/10.1038/nature08821

Ruggiero, M.A., Gordon, D.P., Orrell, T.M., Bailly, N., Bourgoin, T., Brusca, R.C., Cavalier-Smith, T., Guiry, M.D., and Kirk, P.M. (2015). A higher level classification of all living organisms. PLOS ONE *10*, e0119248. https://doi.org/10.1371/journal.pone.0119248

Seemann, T. (2014). Prokka: rapid prokaryotic genome annotation. Bioinformatics *30*, 2068–2069. https://doi.org/10.1093/bioinformatics/btu153

Shapiro, B.J., Friedman, J., Cordero, O.X., Preheim, S.P., Timberlake, S.C., Szabó, G., Polz, M.F., and Alm, E.J. (2012). Population genomics of early events in the ecological differentiation of bacteria. Science *336*, 48–51. https://doi.org/10.1126/science.1218198

Simpson, J.T., Wong, K., Jackman, S.D., Schein, J.E., Jones, S.J., and Birol, I. (2009). ABySS: a parallel assembler for short read sequence data. Genome Res. *19*, 1117–1123. https://doi.org/10.1101/gr.089532.108

Smalley, R., Turcotte, D.L., and Solla, S.A. (1985). A renormalization group approach to the stick-slip behavior of faults. J. Geophys. Res. Solid Earth *90*, 1894–1900.

Stern, A., Bianco, S., Yeh, M.T., Wright, C., Butcher, K., Tang, C., Nielsen, R., and Andino, R. (2014). Costs and benefits of mutational robustness in RNA viruses. Cell Rep. *8*, 1026–1036. https://doi.org/10.1016/j.celrep.2014.07.011

Suerbaum, S., and Josenhans, C. (2007). *Helicobacter pylori* evolution and phenotypic diversification in a changing host. Nat. Rev. Microbiol. *5*, 441–452.

Tang, C., and Bak, P. (1988). Critical exponents and scaling relations for self-organized critical phenomena. Phys. Rev. Lett. *60*, 2347–2350. https://doi.org/10.1103/PhysRevLett.60.2347

Tateno, Y., Imanishi, T., Miyazaki, S., Fukami-Kobayashi, K., Saitou, N., Sugawara, H., and Gojobori, T. (2002). DNA Data Bank of Japan (DDBJ) for genome scale research in life science. Nucleic Acids Res. *30*, 27–30.

Walk, S.T., Alm, E.W., Gordon, D.M., Ram, J.L., Toranzos, G.A., Tiedje, J.M., and Whittam, T.S. (2009). Cryptic lineages of the genus *Escherichia*. Appl. Environ. Microbiol. *75*, 6534–6544. https://doi.org/10.1128/AEM.01262-09

Weimer, B.C. (2012). 100K Pathogen Genome Project,. NCBI Bioproject PRJNA186441.

Weimer, B.C. (2107). 100K Pathogen Genome Project. Genome Announc. *5*, e00594–00517.

Weis, A.M., Storey, D.B., Taff, C.C., Townsend, A.K., Huang, B.C., Kong, N.T., Clothier, K.A., Spinner, A., Byrne, B.A., and Weimer, B.C. (2016). Genomic comparison of *Campylobacter* spp. and their potential for zoonotic transmission between birds, primates, and livestock. Appl. Environ. Microbiol. *82*, 7165–7175.

Weitz, J.S., Hartman, H., and Levin, S.A. (2005). Coevolutionary arms races between bacteria and bacteriophage. Proc. Natl. Acad. Sci. U.S.A. *102*, 9535–9540.

Wilson, E.O. (1999). The Diversity of Life (WW Norton & Company, New York).

Woese, C.R. (2004). A new biology for a new century. Microbiol. Mol. Biol. Rev. *68*, 173–186. https://doi.org/10.1128/MMBR.68.2.173-186.2004

Wu, D., Hugenholtz, P., Mavromatis, K., Pukall, R., Dalin, E., Ivanova, N.N., Kunin, V., Goodwin, L., Wu, M., Tindall, B.J., *et al.* (2009). A phylogeny-driven genomic encyclopaedia of Bacteria and Archaea. Nature *462*, 1056–1060. https://doi.org/10.1038/nature08656

Xiao, Y., Rouzine, I.M., Bianco, S., Acevedo, A., Goldstein, E.F., Farkov, M., Brodsky, L., and Andino, R. (2016). RNA recombination enhances adaptability and is required for virus spread and virulence. Cell Host Microbe *19*, 493–503. https://doi.org/10.1016/j.chom.2016.03.009

Contribution of Metagenomics to Our Understanding of Microbial Processes in Antarctic and Sub-Antarctic Coastal Sediments

5

Mariana Lozada[1*], Hebe M. Dionisi[1], Fernando E. Espínola[1], Priscila A. Calderoli[1], Matías A. Musumeci[1,2], Jessica A. González[1], José L. López[3], Walter P. MacCormack[4,5] and Janet K. Jansson[6]

[1]Laboratorio de Microbiología Ambiental, Centro para el Estudio de Sistemas Marinos (CESIMAR, CONICET), Puerto Madryn, Chubut, Argentina.
[2]Centro de Investigaciones y Transferencia de Entre Ríos (CITER, CONICET), Concordia, Entre Ríos, Argentina.
[3]Cátedra de Virología, Facultad de Farmacia y Bioquímica, Universidad de Buenos Aires, CABA, Argentina.
[4]Instituto Nanobiotec, Facultad de Farmacia y Bioquímica, Universidad de Buenos Aires, CONICET, Buenos Aires, Argentina.
[5]Instituto Antártico Argentino, Buenos Aires, Argentina.
[6]Earth and Biological Sciences Directorate, Pacific Northwest National Laboratory, Richland, WA, USA.

*Correspondence: lozada@cenpat-conicet.gob.ar

https://doi.org/10.21775/9781912530021.05

Abstract
Despite recent advances in metagenomics, increasing our understanding of sediment microbial communities remains a challenge because of their remarkable diversity. In this chapter we describe various strategies for the analysis of metagenomes from complex microbial communities, using a sequence dataset of coastal sediments from Antarctic and sub-Antarctic sheltered environments as a case study. Microbial diversity and community structure were analysed by 16S rRNA gene amplicon sequencing, and shotgun sequencing was used to unveil key metabolic processes at multiple levels. The phylogenetic analysis of biomarker genes (nitrogen cycle, anaerobic hydrocarbon biodegradation) allowed us to shed light on the structure of environmentally relevant, yet low-abundance microbial populations. Gene content and shared synteny in brown algae polysaccharide utilization loci provided evidence

on the evolution of a relevant process for coastal carbon cycling. The statistical analysis of putative microbial rhodopsin sequences showed two levels of phylogeographical segregation, providing evidence on the evolutionary origin of viral sequences. Finally, a workflow was designed for the rational selection of metagenomic sequences encoding enzymes with promising biotechnological features. Each of these analyses provided a glimpse into the biology, ecology and biotechnological potential of these microbial communities, overcoming limitations in coverage and assembly of metagenomes from high-complexity communities.

Introduction

Bacteria constitute the majority of the biomass and carry out most of the metabolic processes in marine sediments, the most widespread habitat on earth (Nealson, 1997; Snelgrove, 1999). Within complex and stratified microbial communities, these organisms have adapted to the utilization of the multiple energy sources and electron acceptors available in the sediments (Snelgrove, 1999). In particular, in coastal environments, sediment bacteria have a key role as essential contributors to the ecosystem services provided by these highly productive ecosystems. Processes carried out by these organisms, such as biogeochemical cycling and pollution control, are critical for the environmental health of coastal systems (Barbier, 2017). Because of their shallow depth, influence from terrestrial sources and exposure to anthropogenic activities, environmental conditions in coastal areas are highly variable, affecting biogeochemical processes and nutrient fluxes (Huettel *et al.*, 2014). This effect is especially relevant in sheltered environments, which present particular hydrography and sediment dynamics, conditions that are relevant for coastal ecology (Klöser *et al.*, 1994).

Coastal regions are increasingly suffering the impact, either locally or globally, of human activities. Land and waterway management, nutrient inputs from agricultural and wastewater discharges, wetland degradation, and climate change are factors likely to increase the degradation of coastal systems over the next decades, leading to dramatic effects on the carbon budget of the oceans (Bauer *et al.*, 2013). In this context, high-latitude coastal environments are regarded as particularly vulnerable. Human populations are continuously moving towards the poles, leading to increased anthropogenic impacts in coastal areas, for example on the Patagonian coast, in South America (Commendatore and Esteves, 2007). Moreover, although Antarctica is still considered one of the most pristine environments in the world, tourism and scientific activities have steadily increased over the past decades, leading to an increment in the impacts on this vulnerable ecosystem (Curtosi *et al.*, 2010). Increase in greenhouse gases and ozone depletion have produced changes in atmospheric circulation patterns, with consequent effects on temperature and sea ice levels, as well as glacier retreatment, mainly in the coastal region of West Antarctica (Turner *et al.*, 2009).

Our fundamental understanding of marine microbial communities, in particular regarding their diversity, structure and function, spatio-temporal variability and the mechanisms that shape them, is still limited. Omic approaches, however, are starting to shed light on the structural and functional properties of these microbial communities, although the large diversity of habitats contained in the global ocean and the inaccessibility of most of these habitats are hindering these studies (Salazar and Sunagawa, 2017). Sampling campaigns focusing on marine environments, such as the International Census of Marine Microbes (Sul *et al.*, 2013), the Global Ocean Sampling (GOS) expedition (Biers *et al.*, 2009), the Tara

Ocean Expedition (Karsenti *et al.*, 2011) and the Ocean Sampling Day (Kopf *et al.*, 2015), are allowing the study of marine microbial communities on an unprecedented geographical scale. However, a bias towards Northern Hemisphere environments must be acknowledged. For instance, according to the Genomes OnLine Database (GOLD; https://gold.jgi.doe.gov, accessed October 2017), out of a total of 26,965 biosamples (defined as the original isolation place where the sample was taken), corresponding to 1078 metagenomic (organism = biome) public studies, only 2185 (< 10%) were from the Southern Hemisphere. In this context, generating microbial community-associated information from high-latitude, remote coastal areas in the southernmost regions of the world is an urgent task, in order to characterize their potential ecological contributions to these vulnerable environments, and to the global ocean ecosystem.

This chapter focuses on the analysis of a metagenomic dataset generated from sediments sampled in two high-latitude coastal environments of the southern hemisphere, characterized by low-energy shores because of their location within sheltered inlets (Fig. 5.1). The first sampling region was Ushuaia Bay, which is a sub-Antarctic environment located within the Beagle Channel, on the south coast of Tierra del Fuego Island, in Argentina (ARG). This channel, which connects the Pacific and Atlantic Oceans in the southernmost tip of South America, was covered by ice until approximately 8000 years ago (Bujalesky, 2007). Ushuaia

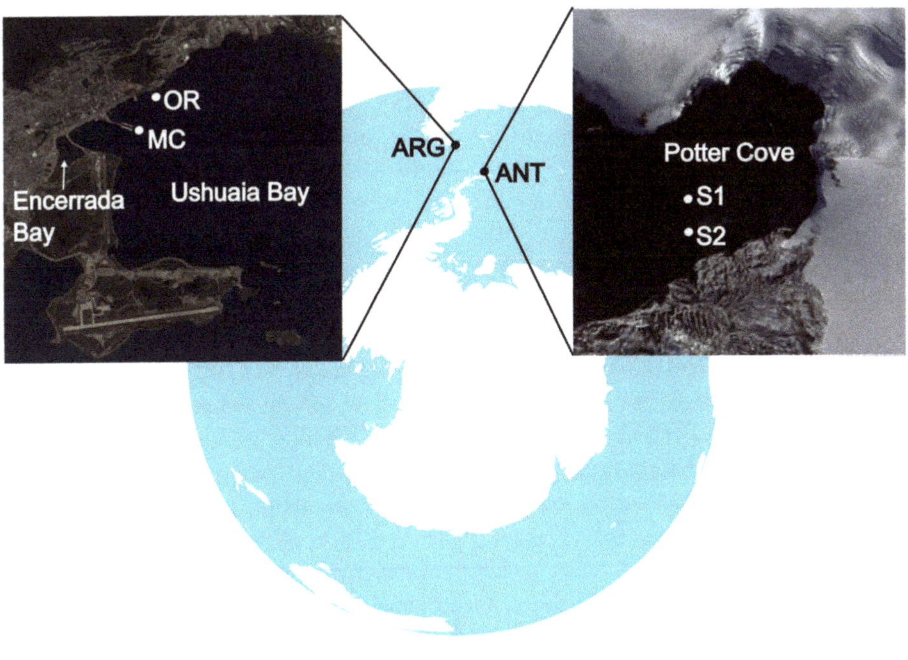

Figure 5.1 Location of the sampling regions and the sites selected within each region. In Ushuaia Bay, Tierra del Fuego, Argentina (ARG): sites MC (54° 48.656′ S 68° 17.731′ W) and OR (54° 48.256′ S 68° 17.296′ W). In Potter Cove, 25 de Mayo (King George) Island, Antarctica (ANT): sites S1 (62° 13.833′ S 58° 39.367′ W) and S2 (62° 13.917′ S 58° 39.300′ W). Satellite images were obtained from Google Earth.

Bay receives seasonal inputs of fresh water from various point sources that produce salinity variations. Ushuaia city, located within this bay, has experienced an explosive demographic growth over the last years, leading to numerous environmental problems associated with urban wastes, eutrophication and oil pollution. In particular, its waters receive the discharge of untreated industrial and domestic wastewater (Dionisi *et al.*, 2011; Gil *et al.*, 2011). An important maritime traffic in Ushuaia Bay (its main port receives around 500 vessels a year; https://www.dpp.gob.ar/web/puerto-ushuaia/estadisticas/evolucion-de-buques/), and a fuel storage facility, constitute sources of chronic hydrocarbon pollution (Commendatore *et al.*, 2012). In addition, metal pollution has been detected in Ushuaia Bay coastal sediments (Duarte *et al.*, 2012; Giarratano *et al.*, 2010). The second sampling region was Potter Cove, located in 25 de Mayo (King George) Island, in the Antarctic Peninsula (ANT) (Fig. 5.1). Potter Cove is a shallow fjord with rapidly changing environmental conditions because of the melting of the Fourcade Glacier. Important seasonal variations also occur in summer as a result of the melting of coastal ice sheets and the thawing of the permafrost, which results in a decrease in salinity and an increase in turbidity in coastal waters (Abele *et al.*, 2017). The Argentinian scientific research station 'Carlini' (previously named 'Jubany') is located near the sampling sites. There have been continuous human activities associated with this base since 1952, and its logistic requirements depend on oil fuels, posing a risk of pollution (Curtosi *et al.*, 2010).

Although both sampling regions are separated by < 1000 km, the Antarctic continent is isolated by the strong Antarctic Circumpolar Current. In fact, evidences of endemicity have been observed in Antarctic microbial communities (Cavicchioli, 2015). One important difference between these two regions is their mean annual seawater temperature, which is 7.0 and −1.1°C for Ushuaia Bay and Esperanza Antarctic Base (150 km south-east of Potter Cove), respectively (Centro Argentino de Datos Oceanográficos, Servicio de Hidrografía Naval, Argentina: http://www.hidro.gob.ar/ceado/Ef/Inventar.asp). These regions have been classified as belonging to different marine ecoregions: Ushuaia belongs to the Magellanic Province, Channels and Fjords of Southern Chile, and 25 de Mayo (King George) Island belongs to Scotia Sea, South Shetland Islands (Spalding *et al.*, 2007). Two sampling sites, approximately 500 m apart, were analysed in each of these sheltered environments. In Ushuaia Bay, the samples were obtained from the Commercial Pier (MC) and the Orion Plant jetty (OR), a small pier connected with a storage facility that is used for fuel loading and offloading (Espínola *et al.*, 2018; Guibert *et al.*, 2016). In Potter Cove, sites corresponded to Station 1 (S1), in the centre of the bay, and Station 2 (S2), near the coast and closer to the scientific base (Fig. 5.1).

The samples were obtained in spring of 2008, when surficial (top 5 cm) subtidal (9–25 m) sediments were collected in triplicate (separated by 3–5 m) at each of the four sampling sites, by scuba diving using cores. A series of environmental variables were obtained *in situ* using multiparameter instruments, and the concentration of hydrocarbons and other organic compounds was determined using gas chromatography–mass spectrometry (Guibert *et al.*, 2016). Sediment water–interface temperature at the time of sampling was 0.2–0.5°C (Potter Cove) and 8.5–8.6°C (Ushuaia Bay). Total hydrocarbon concentrations varied from 0.015 to 0.146 μg/g of wet sediments (ws) in Potter Cove and from 0.069 to 4.29 μg/g ws in Ushuaia Bay. Further details of sampling sites, methods and metadata are available elsewhere (Guibert *et al.*, 2016; López *et al.*, 2017; Matos *et al.*, 2016).

Microbial diversity in cold coastal sediments: insights from large-scale sequencing of 16S rRNA gene amplicons

Despite the environmental relevance of sediment microbial communities for the studied coastal environments, few analyses of their diversity or community structure were available that could allow us to obtain insight into possible lifestyles or ecological roles of their members. In addition, the complexity of sediment communities results in large-scale sequencing being the best (if not the only valid) approach, in order to reach a major fraction of these communities, whose distribution is typically characterized by thousands of microbial types, the majority of them in low abundances (Haegeman *et al.*, 2013). Therefore, we aimed to describe general compositional features of these microbial communities, to highlight differences between both sampled regions, and to detect taxa associated with a particular site (e.g. the oil jetty in Ushuaia Bay). Large-scale sequencing of 16S rRNA gene amplicons was performed as part of the international microbial diversity initiative Earth Microbiome Project (EMP) (Gilbert *et al.*, 2014). This initiative aims to understand the mechanisms possibly involved in shaping the phylogenetic structure of microbial communities in a comparative fashion in various habitats around the world. Habitats such as soils, sediments, water (both saline and non-saline) and host-associated microbiota are sampled under standardized protocols, reducing technical variation and allowing biological patterns to arise (Thompson *et al.*, 2017). New microbial groups are continuously discovered, improving the characterization of the studied environments, and the high number of datasets from diverse samples allows very precise correlation analyses to be performed, enabling the prediction of occurrence patterns of taxa across habitats (Thompson *et al.*, 2017). The Illumina HiSeq high-throughput sequencing platform results in sequence datasets with acceptable coverage values, which is critical in high-diversity environments such as soils or sediments (Caporaso *et al.*, 2012).

Gene fragments were amplified using the primers 515F (5′GTGCCAGCMGCCGCGGTA3′) and 806R (5′GGACTACHVGGGTWTCTAAT3′). This primer set targets the V4 hypervariable region of the ribosomal small subunit (SSU) rRNA gene and has been designed to amplify prokaryotes (bacterial and archaeal 16S rRNA gene), according to standard EMP protocols (http://www.earthmicrobiome.org/emp-standard-protocols/16s/). Preprocessing was performed in QIIME (Caporaso *et al.*, 2010). Briefly, reads were truncated immediately after the first Q3 quality score, and truncated reads that resulted in 75% or less of the original sequence length were eliminated. Sequences with ambiguous bases were also eliminated. Open-reference operational taxonomic unit (OTU) picking was performed at 97% identity threshold in QIIME 1.9.1 against the Greengenes reference database pre-clustered at 97% identity (DeSantis *et al.*, 2006). The open-reference OTU picking is a mixed strategy that combines a fast, reference-based OTU picking step of sequences, with further *de novo* clustering for sequences that did not show hits to references (Rideout *et al.*, 2014). Sequences assigned to chloroplasts and mitochondria were eliminated and the resulting OTU table was subsampled to 134,145 sequences per sample and processed in QIIME to build OTU taxonomic assignment tables and calculate diversity estimators. This step is required in order to analyse all samples in a comparable way, with similar sequencing effort (Gihring *et al.*, 2012). In addition, beta diversity analyses based on relative abundance of taxa (genus and class levels) were performed in R software using the package *vegan*. Briefly, the *OTU* (columns) *per sample* (rows) matrix was used to calculate two-sample similarity indexes (Bray-Curtis, based on relative abundance of shared taxa; Bray and Curtis, 1957) and the resulting *sample* × *sample* similarity matrix was further analysed by hierarchical clustering (UPGMA method). For details, see Guibert *et al.* (2016).

Diversity

Sequence datasets were obtained from four samples from Ushuaia Bay and five from Potter Cove. The unequal size is the result of failure in the amplification/sequencing process (Guibert *et al.*, 2016). Good's coverage values were higher than 97.5% for all samples (Table 5.1).

This allowed a more reliable calculation of diversity estimators, which are all known to be, to a greater or lesser extent, affected by rare species (Haegeman *et al.*, 2013). OTUs were defined at 97% identity. By consensus, OTUs are pragmatically defined at this threshold, as it is considered to correspond roughly to 'species' level. However, it must be noted that there is currently no unambiguous species definition for microbes (Stackebrandt *et al.*, 2002), further complicated by the fact that there is an important variability in distances among taxa at the same phylogenetic level (Schloss and Westcott, 2011). Thousands of OTUs were observed (Table 5.1). Sediments have been shown to be amongst the most diverse environments on earth, and the open reference OTU picking method was fundamental to capture a fraction of the diversity in these environments, which is likely to be composed of divergent and rare taxa (Lozupone and Knight, 2007) and therefore do not readily group with pre-defined references in databases. The Shannon index values estimated for these environments (6–9, Table 5.1) were similar to those found in other marine and coastal sediments, measured with comparable techniques and sequencing effort (Wang *et al.*, 2012). For comparison, in tropical forests, the Shannon index based on plant species is around 3–4 (Lieberman *et al.*, 1996; Naidu and Kumar, 2016). Although diversity was high in all cases, there were significant differences in diversity between regions, with Potter Cove being less diverse and showing more dominance than Ushuaia Bay (Table 5.1). Although richness (Chao1) also showed a decreasing trend in Potter Cove, no significant differences were detected, probably because of data dispersion and a low number of samples (Table 5.1). It must be noted, however, that Chao1 is not an estimator of the true richness but rather of the lowest species richness consistent with the data, and therefore Shannon and related indexes constitute more robust estimators of diversity, as they analyse components of diversity, richness and evenness (relation between relative abundances) (Haegeman *et al.*, 2013).

Table 5.1 Diversity estimators according to 16S RNA gene amplicon sequencing

Region	*n*	*N*	Cov (range)	Obs OTUs (mean ± SD)	Chao1 (mean ± SD)	Shannon* (mean ± SD)	Dominance** (mean ± SD)
Ushuaia Bay (ARG)	4	1.3×10^5	97.9–99.5	6219 ± 2709	8376 ± 3645	9.2 ± 0.4	0.049 ± 0.010
Potter Cove (ANT)	5	1.3×10^5	98.8–99.4	3171 ± 630	4280 ± 1119	6.8 ± 0.5	0.009 ± 0.003

Chao1, Chao1 richness estimator; *n*, number of analyzed samples; Dominance, Simpson's dominance index. *N*, number of sequences per sample (after subsampling); Obs OTUs, observed 97% OTUs; Shannon, Shannon–Wiener diversity estimator. Significant differences between regions are depicted with asterisks (Kruskal–Wallis test, $^*p < 0.05$, $^{**}p < 0.01$).

Community structure

We further aimed to analyse the main taxa composing the microbial communities of the two sampled regions. This was carried out by assigning each 97% OTU a taxonomic label, according to its nearest relative in available reference databases. The Greengenes database was used in this case, currently implemented in QIIME. As OTUs can be defined at whatever level (they are indeed *operational* taxonomic units), the taxonomic information at each level can be utilized for building the community fingerprint. For instance, at the class level, 210 OTUs (taxa) were detected, of which only 23 were higher than 1% in at least one sample (< 1%, Fig. 5.2). This is in agreement with the above measurements of diversity that pointed to the high number of rare taxa in these environments, especially in Ushuaia Bay.

The most abundant classes in the sediments were the Gammaproteobacteria, Deltaproteobacteria (in Ushuaia Bay) and Flavobacteriia (in Potter Cove) (Fig. 5.2A). Among the Gammaproteobacteria from Potter Cove, psychrophilic groups were in high proportion. For instance, the family Psychromonadaceae (mostly assigned to the genus *Psychromonas*) constituted 14.0% ± 6.6% of the total 16S rRNA sequences from the Potter Cove samples, but < 0.1% in Ushuaia Bay. Similarly, sequences assigned to the family Moraxellaceae were 12.4% ± 7.3% of the total 16S rRNA sequences, but were undetected in Ushuaia Bay. The genus *Psychromonas* encompasses various species that are well adapted to marine and other saline aquatic habitats, especially those found in cold environments. Examples are *Psychromonas arctica, Psychromonas antarctica* and *Psychromonas kaikoae*, isolated from cold seawater, Antarctic saline ponds and deep sea sediments, respectively (Groudieva *et al.*, 2003; Mountfort *et al.*, 1998; Nogi *et al.*, 2002). On the other hand, in Ushuaia Bay, two families of Deltaproteobacteria were prevalent: Desulfobacteraceae and Desulfobulbaceae. The sequences assigned to these families constituted 8.4% ± 1.7% and 6.6% ± 3.3% of the sequences, respectively, whereas in Potter Cove these were present in lower abundances (< 1.5% of the reads). Interestingly, the Desulfobacteraceae was highly abundant in the Orion Plant jetty site. Although these abundance values may not seem impressively high, all of them were among the 10 most abundant families at these sites. This is consistent with an abundance distribution of species typical of complex (diverse) communities, which is characterized by the lack of dominant taxa and high number of rare species (Haegeman *et al.*, 2013).

The Flavobacteriia and Alphaproteobacteria (mostly Rhodobacteraceae) were one of the most dominant taxa in Potter Cove. Members of the Flavobacteria group have a pivotal role in the carbon cycle in these environments, participating in the turnover of complex organic matter. Members of this group often have a pool of enzymes specialized in the degradation of algal-derived polysaccharides (Matos *et al.*, 2016) (see section on polysaccharide utilization loci). In addition, the Verrucomicrobiae were very abundant in Antarctic samples (Fig. 5.2A), mostly from the family Verrucomicrobiaceae (Yoon, 2014).

Beta diversity

In accordance with the distinct microbial composition found in each region, clustering analysis using sequences assigned at the class level showed a clear region-based grouping pattern (Fig. 5.2A). Moreover, when OTU level data were used for clustering, samples were grouped by site within each region (Guibert *et al.*, 2016), showing that, despite general composition patterns, site-specific bacterial communities have recently evolved at each site. These more recent events can only be detected with 16S tags at finer (OTU 97%) phylogenetic levels

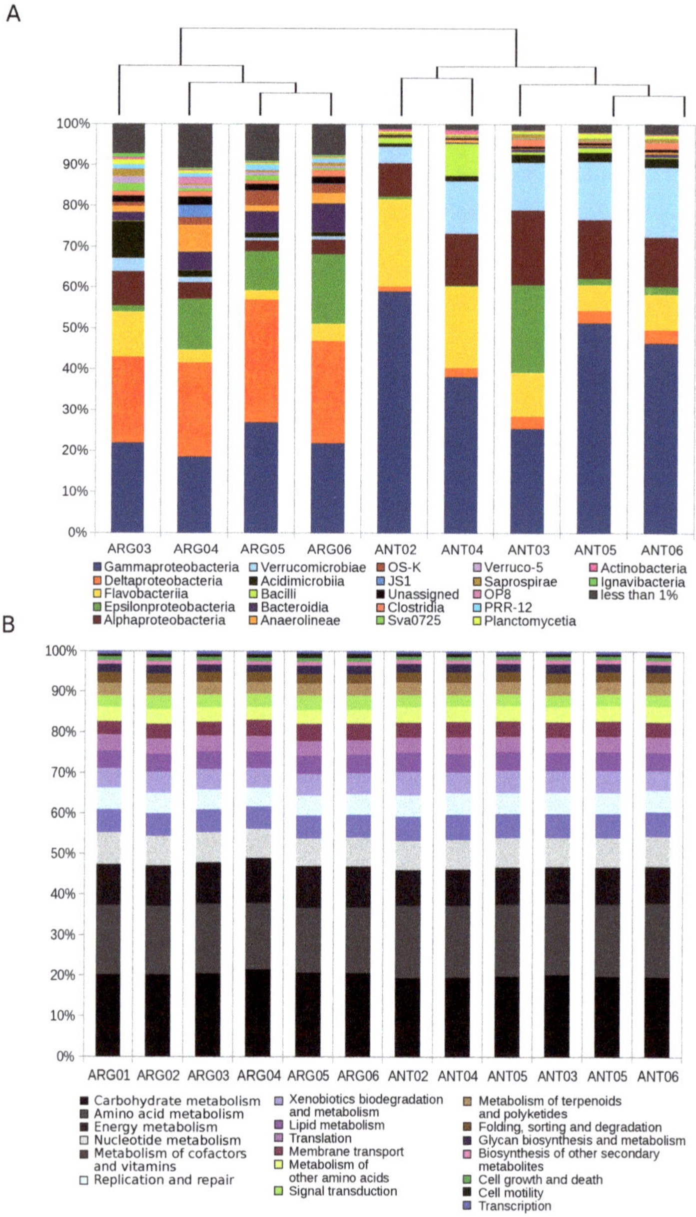

Figure 5.2 General analysis of sediment microbial communities. (A) Community structure at the class level, as estimated by taxonomic assignment of 16S rRNA gene amplicons to a reference database. Clustering analysis was performed using the UPGMA method in R environment. (B) Metabolic potential in the shotgun sequencing dataset from the same metagenomic DNA, using functional annotation of metagenomic sequences to KEGG Orthologous Terms and Pathways. Metagenome accession numbers are as follows: ARG01: 3300000125, ARG02: 3300000131, ARG03: 3300000121, ARG04: 3300000122, ARG05: 3300000242, ARG06: 3300000118, ANT01: 3300000119, ANT02: 3300000136, ANT03: 3300000135, ANT04: 3300000129, ANT05: 3300000132, ANT06: 3300000123 (IMG/M system: https://img.jgi.doe.gov/cgi-bin/m/main.cgi).

(Guibert *et al.*, 2016). This short-term evolutionary changes could be related not only to natural but also to anthropogenic impacts, such as the chronic hydrocarbon pollution found at the oil jetty site in Ushuaia Bay, which has been frequently exposed to small accidental spills over the last few years (Marcos *et al.*, 2012; Guibert *et al.*, 2016).

All of these analyses were implemented with square-root transformed data and using the Bray–Curtis index as a similarity measure, a coefficient that has been widely accepted for use in ecology and environmental sciences. This index calculates the similarity (or dissimilarity, as 1 – value) between community datasets from two samples, taking into account the number of species (or 97% OTUs or other level taxa) shared by two samples, as well as their relative abundance, to construct a similarity value. This coefficient compares two samples (x_1, x_2) in terms of the minimum abundance of each species:

$$\mathrm{SBC}(x_1, x_2) = 2W/(A + B) \tag{5.1}$$

where W is the sum of the minimum abundances of the various species (the abundance at the site where the species is the rarest) and A and B are the sums of the abundances of all species at each of the two sites (Legendre and Legendre, 1998). This index ranges from 0 (no shared species) to 1 (all species are shared and in similar relative abundances). In this way, for example, two samples sharing many OTUs with high relative abundance will have a high Bray–Curtis index, but, if the shared OTUs are very scarce in at least one of the two samples, the index will decrease. This is extremely important because, although not completely quantitative, the abundance information arising from 16S rRNA tags contributes significantly to unveiling patterns of community structure and should be taken into account. In addition, the Bray–Curtis index is insensitive to double zeros, a problem very commonly found in community ecology. Although the presence of a species at two sites is an indicator of their similarity, it is not correct to draw any ecological conclusion from the absence of a species at two sites. Therefore, double zeros should not be taken into account when computing similarity using species presence–absence or abundance data (Legendre and Legendre, 1998). On the other hand, square root transformation has the double advantage of eliminating negative values sometimes arising when this index is transformed to distance, as well as to lighten the contribution of the more abundant species, aiding in the visualization of patterns driven by medium-abundance OTUs. Both data transformations and index choice can have profound effect on the output, allowing possible relations among data to be unveiled when correctly approached (Ramette, 2007).

Analysis of metabolic potential by shotgun sequencing of metagenomic DNA

The analysis of the phylogenetic biomarker 16S rRNA gene by amplicon sequencing provided valuable insights into community structure (diversity and composition) in these coastal environments. The metagenomic approach involving shotgun sequencing of the same samples, using the same extracted DNA but without a previous amplification step, allowed exploration of the metabolic capabilities of the microorganisms inhabiting these assemblages. Sequencing was performed at the Joint Genome Institute (JGI-DOE, USA; https://jgi.doe.gov/) facility using the Illumina HiSeq 2000 platform (2 × 150 bp paired-end reads, one lane per sample). Sequences were obtained from the 12 samples (two regions, two sites in triplicate). The metagenomic dataset contains 101 Gb of sequence data (assembled plus unassembled fractions), with the size of each sample ranging from 2.7 to 16 Gb. Despite the important sequencing effort, the assembly was inefficient, especially in Ushuaia Bay, where an average of only 8.5% of the reads mapped into the scaffolds. This is likely because of the complexity of these microbial communities, characterized by both high richness and evenness (see above results on diversity). However, this did not prevent ecological analysis, as both assembled and unassembled fractions were used as information sources. The metagenomes were annotated and analysed through the Integrated Microbial Genomes with Microbiome Samples (IMG/M) (Huntemann *et al.*, 2016). The metagenomes are available at the IMG/M system (https://img.jgi.doe.gov/cgi-bin/m/main.cgi), under accession numbers ARG01: 3300000125, ARG02: 3300000131, ARG03: 3300000121, ARG04: 3300000122, ARG05: 3300000242, ARG06: 3300000118, ANT01: 3300000119, ANT02: 3300000136, ANT03: 3300000135, ANT04: 3300000129, ANT05: 3300000132 and ANT06: 3300000123. Assignment to KEGG Orthology (KO) Terms and Pathways (Kanehisa, 2002) was used as functional evidence to analyse the metagenomes in terms of their potential functions.

The estimated gene copies (as defined by IMG/M, which corrects for read depth in the assembled fraction) of coding sequences assigned to the 17,948 KOs available at the system were downloaded and mapped into KEGG pathways and modules (level 2: general categories; level 3: pathways; level 4: modules) and normalized to avoid biases resulting from different metagenome sizes (Espínola *et al.*, 2018). The most abundant functional categories found in the metagenomes were related to carbohydrate, amino acid and energy metabolisms. It is noteworthy that xenobiotic degradation and metabolism of terpenoids and polyketides were also well represented, highlighting the potential role in environmental depuration of sediment microbial communities (Fig. 5.2B). In contrast, with phylogenetic structure, which was different between the two regions and even variable among sites, the metabolic potential of metagenomes was strikingly similar. This result was also observed with a larger dataset including samples of similar environments from the Northern Hemisphere (Espínola *et al.*, 2018). Moreover, a comparative analysis was performed with metagenome datasets from other environments and evidence was found of a higher similarity of sediment metagenomes with other high-complexity environments (marine sediments, freshwater sediments and soils) (Espínola *et al.*, 2018). These results suggest that phylogenetically distinct microbial communities have adapted to perform similar core metabolic functions according to their environmental constraints, as has been previously proposed (Acosta-González *et al.*, 2013; Muegge *et al.*, 2011). Moreover, they show that habitat and biological complexity play a fundamental role in the functional structuring of microbial assemblages. Despite these similarities, a number of locally defined traits, as well as traits varying with specific environmental factors, could be observed when analysing the metagenomic data at a finer scale (Espínola *et al.*, 2018).

In addition to inferences on general metabolic potential, the analysis of metagenomes by shotgun sequencing allowed us to obtain insights into the characteristics of certain microbial groups that were revealed to be abundant in these environments by amplicon sequencing. For instance, we analysed the scaffolds assigned to the genus *Psychromonas*, whose 16S rRNA sequences were prevalent in Potter Cove samples (see above). A total of 409 scaffolds longer than 10 kb assigned to *Psychromonas* by the IMG/M functional annotation pipeline were retrieved from Potter Cove metagenomic data. These scaffolds contained a total of

6760 coding sequences and summed to a total of 6.5 Mb of sequence data from these microbial populations, equivalent to the size of a *Psychromonas* genome (Mukherjee *et al.*, 2017). Annotation of the amino acid sequences of these scaffolds through the GhostKOALA automatic annotation server (Kanehisa *et al.*, 2016) showed a wide array of functions potentially encoded: 85 out of 179 described pathway modules (47%) were found to be complete, including the capability for dissimilatory nitrate reduction to ammonia, various modules comprising the metabolism of carbohydrates, lipids, nucleotides, amino acids, cofactors and vitamins, and a number of transport and two-component regulatory systems (phosphate starvation and protein misfolding responses, anoxic redox control, copper resistance, chemotaxis, nitrogen regulation and nitrate respiration, etc.). These results indicate that important metabolic information about Antarctic *Psychromonas* species could be retrieved by attaining genome reconstruction from these metagenomes. In fact, the metagenomes from Potter Cove were found to be the top source of *Psychromonas* sequences, as revealed in a global scale analysis by recruitment of coding sequences from metagenomes with reference isolate genomes [Genomic Encyclopedia of Bacteria and Archaea (GEBA) initiative] (Mukherjee *et al.*, 2017). It must be noted, however, that the metagenomic sequences obtained by this method may belong to more than one population, in part because of the way in which the taxonomic assignment of metagenomic sequences is carried out (by recruitment against reference genomes), further complicated by their divergent nature.

Analysis of biomarker genes for processes of environmental relevance

Anaerobic hydrocarbon biodegradation

As described above, the studied sites have a certain level of anthropogenic impact because of the different human activities that are usually carried out in these environments. In Potter Cove, the analysis of organic markers revealed that oil hydrocarbons and sewage inputs were positioned in a range of low values (Dauner *et al.*, 2015); in Ushuaia Bay, both sites showed moderate levels of hydrocarbon pollution, likely from maritime traffic (MC, the commercial pier) and/or as a result of the spills from an oil jetty (OR, Orion Plant). The Orion Plant site in particular has been studied in detail throughout the last years. Hydrocarbon levels were determined in both intertidal and subtidal sediments and results were coupled to characterization of microbial communities by molecular techniques. Samples from 2004 to 2010 were analysed. Aliphatic hydrocarbons were found to be in the range of 0.27–77 μg/g of sediment and polycyclic aromatic hydrocarbons (PAHs) were in the range from 0.1 to 4.7 μg/g (Espínola *et al.*, 2018; Guibert *et al.*, 2016; Marcos *et al.*, 2012). Various hydrocarbon degradation genes were amplified by polymerase chain reaction (PCR) with different primers, using DNA from sediments of this site as a template. Archetypical (e.g. similar to the obligate PAH-degrading genus *Cycloclasticus*) as well as novel, divergent variants of PAH-degradation genes were described in these works (Espínola *et al.*, 2018; Guibert *et al.*, 2016; Marcos *et al.*, 2012). Alkane degradation genes (AlkB-like alkane monooxygenases; Nie *et al.*, 2014; van Beilen and Funhoff, 2007) were also very abundant and diverse (Espínola *et al.*, 2018; Guibert *et al.*, 2012). Moreover, a response was observed to the addition of both crude oil and individual PAHs, which was detected at the community level by large-scale sequencing of 16S rRNA gene amplicons (Guibert *et al.*, 2012) and by targeting specific populations by quantitative polymerase chain reaction (qPCR) analysis of functional biomarker genes (Loviso *et al.*, 2015). These studies showed that a strongly adapted microbial community has evolved at this chronically polluted site, with a genetic potential that makes it capable of depuration of a variety of oil-derived organic compounds, including carcinogenic molecules such as pyrene (https://cameochemicals.noaa.gov/chemical/5163) (Loviso *et al.*, 2015).

Oil biodegradation is a complex process involving a very high number of microbial populations, the majority of them uncultured, which are characterized by biotic and abiotic interactions (Cravo-Laureau and Duran, 2014). Although much was advanced in our understanding of hydrocarbon-degrading communities in cold coastal sediments, only a limited insight of their potential capabilities was obtained, as the analysed genes coded for enzymes catalysing aerobic hydrocarbon degradation. Sediments are characterized by steep redox gradients and quickly become anoxic with depth, especially in subtidal environments and/or with high organic matter load. In these conditions, nitrate or sulfate rather than oxygen are the main electron acceptors for respiration, and anaerobic hydrocarbon degradation is therefore expected to occur (Muegge *et al.*, 2011). The genes for anaerobic degradation processes are less well known than their aerobic counterparts as regards their diversity in the environment, and for many of them the only available information is from pure cultures or enrichments (Rabus *et al.*, 2016). In a recent work involving the metagenomic analysis of subtidal coastal sediments from cold environments, including the ones described in this chapter, 14 hydrocarbon degradation biomarker genes were analysed. These biomarkers code for enzymes involved in the degradation of PAHs, benzene, toluene, ethylbenzene and xylene (BTEX) and alkanes. From this analysis, it could be observed that protein-coding sequences assigned to anaerobic hydrocarbon degradation biomarkers (BssA, BamA, EbdA, BcrA) were the second most abundant group, only outnumbered by AlkB (encoding alkane-1-monooxygenase) (Espínola *et al.*, 2018). AlkB is normally encountered in chronically polluted environments, attributed to the fact that alkanes are the most readily degradable portion of the oil mixture (Bargiela *et al.*, 2015; Head *et al.*, 2006). The structure of the community with hydrocarbon-degrading potential, as revealed by functional biomarkers, suggested that anaerobic biodegradation was a process of environmental relevance in the sediments.

We further focused on the gene encoding benzylsuccinate synthase (BSS) and its related enzymes, for which more detailed information has been obtained over the last years. This is a family of enzymes that catalyse a number of reactions, all of which involve fumarate addition to different substrates. The large subunit of this enzyme belongs to the pyruvate formate lyase (PFL) family of glycyl radical enzymes, and as it contains the active site it is currently used as a functional biomarker for anaerobic hydrocarbon biodegradation (Acosta-González *et al.*, 2013; Callaghan *et al.*, 2010). Some members of this family include archetypical BssA (benzylsuccinate synthase, catalysing benzene degradation), AssA (alkylsuccinate synthase, for alkane degradation) and NmsA (naphthyl-2-methylsuccinate synthase, participating in methylnapthalene degradation). Notably, because of their evolutionary history, the sequences from these genes are phylogenetically clustered by substrate and further by type of environment (marine, terrestrial origin), leading to the possibility of analysing environmental sequences and making inferences on their possible substrate and ecological role (Callaghan *et al.*, 2010). Interestingly, the number of sequences assigned to this biomarker correlated with hydrocarbon concentrations in sediments (Espínola *et al.*, 2018).

We analysed the sequences assigned to BssA and related enzymes in the sediments from Ushuaia Bay and Potter Cove, based on the functional annotation of metagenomic sequences to KO terms. The sequences assigned to the KO of interest [K07540, corresponding to 'benzylsuccinate synthase' (EC 4.1.99.11)], were obtained from the shotgun sequencing dataset. For this analysis, the corresponding amino acid sequences were downloaded from the IMG website (using 'function search' to search by KO ID and selecting the results for export in *fasta* format). Although only 70 sequences were obtained from Potter Cove metagenomes, 1315 sequences were obtained from Ushuaia Bay, which were analysed further. These numbers were in accordance with the higher relative abundance of this biomarker gene sequences in this region, especially in the Orion Plant site (Espínola *et al.*, 2018). For the phylogenetic assignment of Ushuaia Bay environmental sequences, an evolutionary placement algorithm (EPA) was carried out (Berger *et al.*, 2011). This analysis performs the independent assignment of short sequences on to the edges of a fixed reference tree constructed from full-length sequences, under the maximum likelihood model (Berger *et al.*, 2011; Stamatakis, 2014). The likelihood of the placements along the tree is calculated for each sequence and the phylogenetic assignment is given based on the best placement. This analysis is especially suitable for metagenomic sequences, whose size distribution is biased towards short fragments, and where a multiple alignment is not possible as they do not overlap (in contrast to the biomarker data generated by PCR amplification). Ushuaia Bay sequences ranged from 25 to 94 amino acids in length. A reference alignment was built using 122 pure culture and environmental sequences including representatives from all BssA-like homologues. The alignment was built in ClustalX (Thompson *et al.*, 2002), a hidden Markov model was built from the alignment in HMMER 3.1b1 (Eddy, 2009) using *hmmuild* and the metagenomic sequences were further placed on to this alignment using *hmmalign*. A reference tree was constructed with RaxML (Stamatakis, 2014), using a previously assessed evolutionary model (Espínola *et al.*, 2018). The metagenomic sequences were then placed into the reference tree using EPA in RAxML. This analysis produces a number of files, including the classification table with the maximum likelihood values and the edges of the tree to which the sequences were assigned, as well as the labelled reference tree with the edge information. Processing this information (e.g. in R software environment) allows the extraction of the number of sequences assigned to each cluster of the tree. In addition, for visualization, a tree was constructed by using the 'portable tree' (.jplace file) provided by EPA-RAxML, with the reference and metagenomic sequences information. This was done by using the 'fat' command in *guppy* software (Matsen *et al.*, 2010), which generates a tree in which branch widths are proportional to the amount of sequences assigned to the corresponding branch. The tree was visualized in Archaeopteryx (Han and Zmasek, 2009).

The results are shown in Fig. 5.3. Of the 1315 sequences, 81% (1066) could be classified by this method, being assigned to a specific cluster in the tree. The most abundant variant was AssA, which corresponded to 67.5% of the assigned sequences, which also included sequences assigned to MasD (methylalkyl succinate synthase, for branched compounds) (see branch widths in this cluster in Fig. 5.3). These variants were followed in abundance by archetypical BssA, which represented 12.4% of the sequences (toluene-specific BssA and BTEX including p-xylene clusters) (Callaghan et al., 2010). Interestingly, 9.9% of the sequences were found to be related to hydroxybenzylsuccinate synthase (HbsA, catalysing p-cresol degradation) and other genes of unknown function from *Desulfobacula toluolica* (Desulfobacteraceae), a marine, sulfate-reducer Deltaproteobacteria that is able

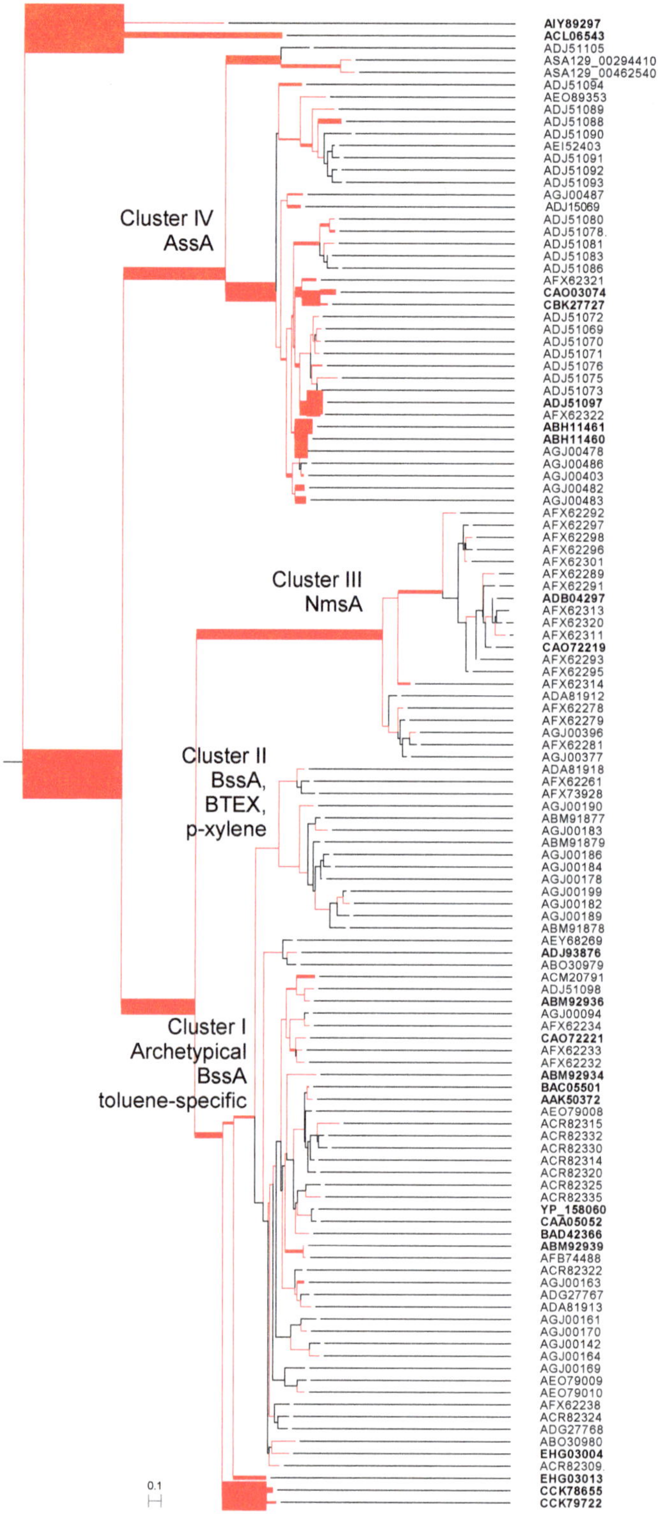
AIY89297
ACL06543
ADJ51105
ASA129_00294410
ASA129_00462540
ADJ51094
AEO89353
ADJ51089
ADJ51088
ADJ51090
AEI52403
ADJ51091
ADJ51092
ADJ51093
AGJ00487
ADJ15069
ADJ51080
ADJ51078
ADJ51081
ADJ51083
ADJ51086
AFX62321
CAO03074
CBK27727
ADJ51072
ADJ51069
ADJ51070
ADJ51071
ADJ51076
ADJ51075
ADJ51073
ADJ51097
AFX62322
ABH11461
ABH11460
AGJ00478
AGJ00486
AGJ00403
AGJ00482
AGJ00483
AFX62292
AFX62297
AFX62298
AFX62296
AFX62301
AFX62289
AFX62291
ADB04297
AFX62313
AFX62320
AFX62311
CAO72219
AFX62293
AFX62295
AFX62314
ADA81912
AFX62278
AFX62279
AGJ00396
AFX62281
AGJ00377
ADA81918
AFX62261
AFX73928
AGJ00190
ABM91877
AGJ00183
ABM91879
AGJ00186
AGJ00184
AGJ00178
AGJ00199
AGJ00182
AGJ00189
ABM91878
AEY68269
ADJ93876
ABO30979
ACM20791
ADJ51098
ABM92936
AGJ00094
AFX62234
CAO72221
AFX62233
AFX62232
ABM92934
BAC05501
AAK50372
AEO79008
ACR82315
ACR82332
ACR82330
ACR82314
ACR82320
ACR82325
ACR82335
YP_158060
CAA05052
BAD42366
ABM92939
AFB74488
ACR82322
AGJ00163
ADG27767
ADA81913
AGJ00161
AGJ00170
AGJ00142
AGJ00164
AGJ00169
AEO79009
AEO79010
AFX62238
ACR82324
ADG27768
ABO30980
EHG03004
ACR82309
EHG03013
CCK78655
CCK79722
Cluster IV
AssA
Cluster III
NmsA
Cluster II
BssA,
BTEX,
p-xylene
Cluster I
Archetypical
BssA
toluene-specific
0.1

Figure 5.3 (see p. 78) Phylogenetic placement of metagenomic sequences assigned to K07540 ('BssA') in Ushuaia Bay metagenomes. Numbers on the right correspond to accession numbers in the GenBank Protein Database. Accession numbers for reference sequences are depicted in bold. Complete information (accession number: protein name, organism name): ACL06543.1: Pfl *Desulfatibacillum alkenivorans* AK-1; AAK50372: BssA *Azoarcus* sp. T; BAC05501: BssA *Thauera* sp. DNT-1; CAA05052: BssA *Thauera aromatica*; AIY89297: AssA/MasD *Geoglobus acetivorans*; BAD42366: putative BssA *Magnetospirillum* sp. TS-6; ABH11460: AssA, sulfate reducing bacterium AK-1; ABM92939: BssA partial sulfate reducing bacterium TRM1; CBK27727: MasA (1-methylalkyl) succinate synthase *Aromatoleum* sp. OcN1; ADJ51097: AssA *Desulfoglaeba alkanexedens* ALDC; ADJ93876: putative benzylsuccinate synthase, Clostridia bacterium enrichment culture clone BF; ADB04297: putative NmsA (naphthyl-2-methyl succinate synthase), partial bacterium enrichment culture clone N47; CAO72219: alpha subunit of naphtylmethylsuccinate synthase delta proteobacterium NaphS2; YP158060: BssA *Aromatoleum aromaticum* EbN1; ABM92934: BssA partial *Desulfosarcina cetonica*; ABM92936: BssA partial *Geobacter grbiciae*; CAO72221: BssA *Desulfobacula toluolica* Tol2; EHG03004.1: Pfl *Desulfotomaculum gibsoniae* DSM 7213; CCK78655: HbsA hydroxybenzylsuccinate synthase alpha subunit *Desulfobacula toluolica* Tol2; CCK79722: BssA3 benzylsuccinate synthase alpha subunit *Desulfobacula toluolica* Tol2; ABH11461: alkylsuccinate synthase, sulfate reducing bacterium AK-1; CAO03074: 1-methyl alkyl succinate synthase subunit MasD, *Azoarcus* sp. HxN1; EHG03013: Pfl PFL *Desulfotomaculum gibsoniae* DSM 7213. The other accession numbers (not depicted in bold) correspond to environmental sequences, also used as references. The analysis was performed in RAxML using the EPA.

to degrade a wide range of aromatic compounds (Wöhlbrand *et al.*, 2013). According to previous works (Callaghan *et al.*, 2010), BssA-like sequence analysis supports separation between marine and terrestrial origins and further subgrouping by taxon, suggesting that the metagenomic sequences described in this chapter could correspond to typically marine variants. Sequences related to NmsA were also detected in the sediments (6.3%) (Fig. 5.3). These results indicate that there is an extended potential for the degradation of a wide range of compounds in these sediments, including linear and branched aliphatic hydrocarbons, BTEX and PAHs.

Potential for cycling and depuration of toxic forms of nitrogen

Coastal regions are vulnerable to different sources of nitrogen inputs by anthropogenic activities and physical processes that can cause eutrophication (Herbert, 1999). Nitrogen (N) turnover is only driven by microbial activity and therefore microbial communities constitute an extremely relevant variable in nutrient-loaded coastal ecosystems. The N-cycle includes several chemical transformations carried out by specific microbial groups and includes the following metabolic pathways: N-fixation, ammonification, nitrate assimilation, nitrification, denitrification, respiratory ammonification (or dissimilatory nitrate reduction to ammonia, DNRA) and anaerobic ammonium oxidation (anammox) (Hallin *et al.*, 2017). The N-cycle has been widely studied in marine environments by different approaches and methodologies, allowing the continuous discovery of new microorganisms as well as entirely new pathways (Zehr and Kudela, 2011). Particularly, over the last two decades, our understanding of the nitrification process in shallow coastal sediments has been broadened, allowing the recognition of its importance. Nitrification plays an important role in the coastal N-cycle, catalysing the oxidation of ammonium (NH_4^+) released from the decomposition of organic matter. The resulting nitrate is available for primary production, which in excess and under certain environmental conditions can be the cause of eutrophication (Zehr and Kudela, 2011). In addition, the coupling of nitrification and denitrification processes generates reduced forms of N: nitric oxide (NO), nitrous oxide (N_2O) and dinitrogen (N_2). These N species are released as gases into the atmosphere, contributing to the removal of excess nutrients in coastal areas. However, it must be noted that N_2O is a very powerful ozone-depleting gas, raising environmental concerns regarding the extent to which denitrification occurs and the optimal conditions for the occurrence of a complete denitrification process, the reduction of nitrate (NO_3^-) to N_2 (Ravishankara *et al.*, 2009).

Nitrification is a two-step process. The first step involves two reactions. The first reaction is catalyzed by the ammonia monooxygenase (AMO), with NH_4^+ oxidized to nitrite (NO_2^-) via the intermediate hydroxylamine (NH_2OH) (equation 5.2). The oxidation of NH_2OH follows rapidly, catalyzed by the activity of hydroxylamine oxidoreductase (HAO) (equation 5.3). This is the key reaction by which aerobic ammonia oxidizers produce energy for cell growth, coupled to the activity of two c-type cytochromes and funneling to the quinone system (Klotz and Stein, 2008). Finally, NO_2^- is oxidized to NO_3^- by the nitrite oxidoreductase action (equation 5.4). During nitrification, N_2O can be produced as a by-product through the oxidation of NH_2OH (nitrification-dependent N_2O production) or through the aerobic reduction of NO_2^- (nitrifier denitrification) (Stein and Klotz, 2011).

$$\mathbf{NH_3} + O_2 + 2H^+ + 2e^- \rightarrow NH_2OH + H_2O \text{ (ammonia monooxygenase)} \tag{5.2}$$

$$NH_2OH + H_2O \rightarrow \mathbf{NO_2^-} + 5H^+ + 4e^- \text{ (hydroxylamine oxidoreductase)} \tag{5.3}$$

$$\mathbf{NO_2^-} + H_2O \rightarrow \mathbf{NO_3^-} + 2H^+ \text{ (nitrite oxidoreductase)} \tag{5.4}$$

The two steps of nitrification are catalysed by different microbial clades. The aerobic oxidation of NH_4^+ to NO_2^- is carried out by ammonia-oxidizing bacteria (AOB), whereas the oxidation of NO_2^- to NO_3^- is catalysed by nitrite-oxidizing bacteria (NOB). Although originally, the only representatives known for these respective groups were the genera *Nitrosomonas* and *Nitrobacter*, over the last two decades, the use of culture-independent methods has resulted in an increase in our knowledge about the diversity of ammonia-oxidizing organisms in different environments, revealing that it is far from complete (Figuerola and Erijman, 2010; Gieseke *et al.*, 2003; Head *et al.*, 1993). Moreover, the analysis of metagenomes from engineered systems and *in situ* experiments revealed that members of the *Nitrospira* are capable of carrying out the complete nitrification process, changing a long-held paradigm (Daims *et al.*, 2015). On the other hand, the finding of bacteria capable of anaerobic oxidation of ammonium to nitrogen (anammox process; Strous *et al.*, 1999), and of a marine aerobic ammonium-oxidizing archaeon (AOA; Könneke *et al.*, 2005), revealed the complexity of ammonium oxidation processes and the diversity of the associated microbiota. The AOB and AOA contain the *amo* operon that consists of at least three genes, *amo*C, *amo*A and *amo*B (Norton *et al.*, 2002). In particular, *amo*A encodes the subunit containing the putative enzyme active site and has been extensively used as a biomarker of ammonia-oxidizing microorganisms in natural environments (Bothe *et al.*, 2000). In addition, *amo*B has also been proposed as a alternative candidate for phylogenetic studies because of its suitable size (ca. 1250 bp) and its essential role in the process (Calvó and Garcia-Gil, 2004). In AOB, the conversion of hydroxylamine to NO_2^- is catalysed by a periplasmic enzyme codified by the *hao* gene (Bergmann *et al.*, 2005). In AOA, in contrast, no homologue of this enzyme has been yet identified (Reyes *et al.*, 2017).

The environmental status of Ushuaia Bay is comparable to an eutrophicated system, due in part to the explosive population growth that occurred in Ushuaia city over the last few years (see 'Introduction'). This urbanization process was accompanied by a constant discharge of untreated urban and industrial effluents into the low-energy shores of Ushuaia Bay, with the consequent raise in ammonium and organic matter levels (Gil *et al.*, 2011), a condition that is likely to affect the structure of the portion of the community capable of N-cycling (N-cycling *guild*). A biomarker-based metagenomic survey was carried out to uncover the potential for nitrification in Ushuaia Bay and Potter Cove metagenomes. This approach overcomes the limitation associated with the culture of nitrifying bacteria, which are recalcitrant to isolation and growth in laboratory conditions. Another advantage of this approach is that it does not rely on the use of primers, which probably suffer important biases because of our still poor knowledge of the diversity of nitrifying phylotypes (Stein, 2011).

We focused on deduced amino acid sequences assigned to three biomarker genes, coding for enzymes involved in ammonium oxidation: AmoA (ammonia monooxygenase α-subunit), AmoB (ammonia monooxygenase β-subunit) and Hao (hydroxylamine oxidoreductase) (Calvó and Garcia-Gil, 2004; Rotthauwe *et al.*, 1997). For this purpose, associated KO identifiers (K10944 for *amo*A, K10945 for *amo*B and K10535 for *hao*) were used as functional evidence for searches in the metagenomic dataset. The estimated relative abundance of each biomarker was calculated by dividing the number of amino acid sequences assigned to each KO by the total number of amino acid sequences assigned to KOs in each metagenome (for details, see previous section on hydrocarbon degradation biomarker genes). For comparisons, 121 metagenomes representative of different habitats including algae, seawater, freshwater, freshwater and marine sediments, soil and the human body microbiome were selected from IMG/M (Espínola *et al.*, 2018) and analyzed in the same way.

Nitrification biomarkers were found at low relative abundance: together, they accounted for < 0.15% and 0.03% of the total sequences assigned to KOs in Ushuaia Bay and Potter Cove, respectively. This result was not surprising, as nitrogen metabolism usually represents < 1% of the total number of sequences assigned to hierarchical metabolisms in metagenomes obtained from highly diverse environments such as soils and sediments (Delmont *et al.*, 2011; Mason *et al.*, 2014; Reyes *et al.*, 2017; Sessitsch *et al.*, 2012). In addition, within the sequences related to the N-cycle, the ones associated with N-fixation and nitrification are usually found to be underrepresented with respect to other subprocesses such as denitrification and DNRA (Delmont *et al.*, 2011; Mason *et al.*, 2014; Reyes *et al.*, 2017; Yu and Zhang, 2012). Low abundance, however, does not necessarily mean that the process is not relevant for the ecosystem as a whole. Very important ecosystem functions, such as nutrient cycling (e.g N-fixation, nitrification, methanogenesis), pathogenesis and metabolite production, are often carried out by low-abundance microbial groups ('keystone species'; Riesenfeld *et al.*, 2004).

It is noteworthy that the proportion of sequences assigned to Hao (0.024 ± 0.007% in Ushuaia Bay and 0.004 ± 0.001% in Potter Cove) was significantly higher than that assigned to AmoA/B (0.0004 ± 0.0001% in Ushuaia Bay and 0.0002 ± 0.0001% in Potter Cove) (Welch two-sample test, p-value = 0.001). Especially in Ushuaia Bay, Hao sequences were 32 times more abundant than those coding for AmoA/B. Moreover, sequences assigned to Amo were underrepresented in these cold sediments with respect to the average in metagenomes from various environments (data not shown). This result may be the result of differences in the copy number of these genes. For instance, it has been observed at least in archetypical nitrifying microorganisms (e.g. *Nitrosomonas europaea*) that nitrification genes can be present in multiple copies (Hommes *et al.*, 2001). The fact that AmoA and AmoB were in approximately the same relative abundance supports this hypothesis. Alternatively, it could be the result of problems in the process of functional annotation of these specific genes, as these metagenomes were characterized by the divergence of their sequences with respect to those deposited in databases (Espínola *et al.*, 2018; Matos *et al.*, 2016).

Sequences assigned to Hao were significantly more abundant in Ushuaia Bay than in Potter Cove. Moreover, they were approximately six times more abundant in Ushuaia Bay than the general average across environments (0.024% ± 0.007% in Ushuaia Bay vs. 0.004% ± 0.007% average across 121 metagenomes from different habitats). In sediments from Potter Cove, in contrast, no overrepresentation was observed for Hao. In order to assess the structure of the putative nitrification guild carrying this biomarker, the metagenomic sequences assigned to Hao were analysed phylogenetically using the EPA in RaxML (see method description in the previous section). For Hao sequences, the reference alignment and tree were constructed using the sequences from Schmid *et al.* (2008). This reference dataset includes diverse representatives of Hao as well as their homologue Hzo (hydrazine oxidoreductase). This enzyme has been described in anammox bacteria, participating in the oxidation of hydrazine (N_2H_4) and generating dinitrogen gas (N_2) as the final product (Strous *et al.*, 2006).

From this analysis, it can be observed that the major clusters are well supported, and monophyletic clades are formed for Hao and Hzo, allowing the further placement of metagenomic sequences assigned to K10535 in the different clusters of the tree (Fig. 5.4A).

The identification of the position of metagenomic sequences along the different branches of the tree enables not only their classification as Hao or Hzo but also the inference of their phylogenetic origin (e.g. Betaproteobacteria-like Hao, if they fall in the corresponding branch of Fig. 5.4A). This is performed through the processing of the information obtained as a result of EPA analysis (for details see previous section).

A total of 19,752 sequences assigned to K10535 (15,676 from Ushuaia Bay and 4076 from Potter Cove) were phylogenetically placed with this method. Despite the differences in the relative abundance of this biomarker between both regions (Fig. 5.4B, upper panel), the structure of the nitrification-associated community was similar in all cases (Fig. 5.4B, lower panel). A high proportion of sequences per sample (36 ± 3%) could not be further classified, which could be because of their short size (average length was 50 amino acids). In addition, this analysis revealed that an important fraction of sequences (24 ± 1%) was indeed phylogenetically placed in the Hzo cluster, unveiling the presence of biomarkers for the anammox process, as well as an overestimation of metagenomic sequences assigned to Hao (and of nitrification potential) when using K10535 as functional evidence.

Among the sequences that were positioned in the Hao cluster, which accounted for 29 ± 6% of the total sequences assigned to this KO, the majority were associated with the well-characterized Betaproteobacteria group (genera *Nitrosomonas* and *Nitrosospira*)

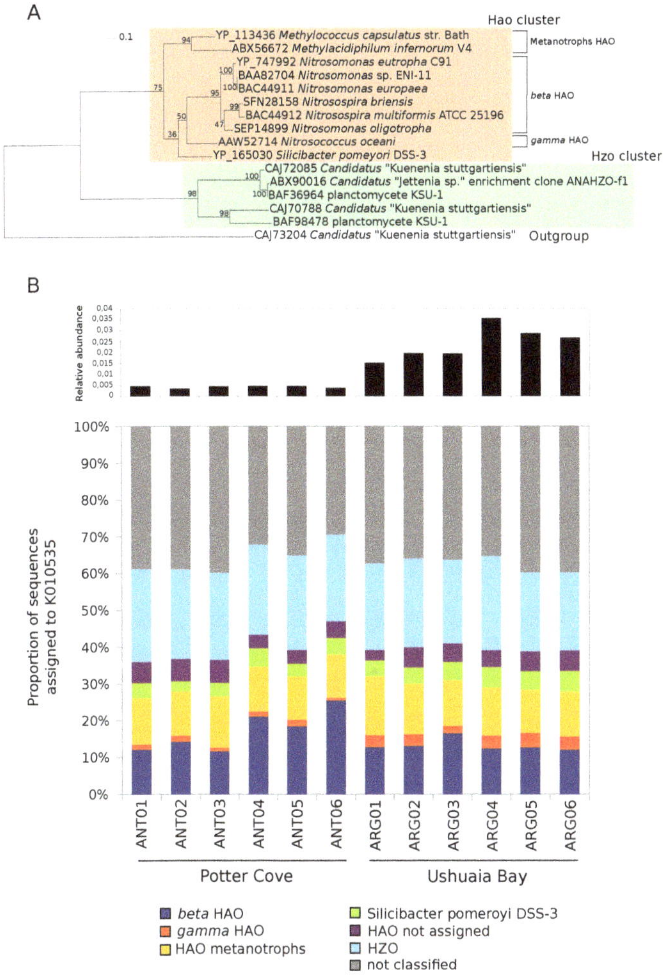

Figure 5.4 Diversity of nitrification biomarkers in sediment metagenomes from Ushuaia Bay and Potter Cove. Amino acid sequences assigned to K010535 ('Hao') were analysed through the EPA. (A) Phylogenetic tree of reference sequences, constructed by maximum likelihood in RAxML. Reference sequences are indicated by GenBank accession numbers followed by organism name. (B) Upper panel: relative abundance of amino acid sequences assigned to K010535 in Ushuaia Bay and Potter Cove, discriminated by sample. Lower panel: classification of the metagenomic sequences according to their position in the reference tree as retrieved by the EPA. The bars are stacked to 100% with regard to the total amino acid sequences assigned to K010535 ('Hao') in each metagenome.

(15% ± 4%). Another well-represented Hao group was the one formed by sequences described in methanotrophic bacteria (13% ± 1%) (Fig. 5.4B). This microbial group is aerobic and obtains carbon and energy from the oxidation of methane. It is widely distributed in various environments, playing a role in the detoxification of this potent greenhouse gas (Lidstrom, 2006). As the particulate methane monooxygenase (pMMO) is homologous to ammonium monooxygenase (Campbell *et al.*, 2011; Klotz *et al.*, 2008), methanotrophic bacteria are capable of carrying out nitrification. Moreover, ammonium competes with methane in the enzymatic reaction, giving hydroxylamine as a product (Stein and Klotz, 2011). Hydroxylamine is a toxic intermediate and therefore it is usual to find *hao* genes in methane-oxidizing bacteria. In contrast to what happens in archetypical nitrifiers, in this case oxidation of hydroxylamine is not coupled to cell growth (Stein and Klotz, 2011). Interestingly, although the HAO enzyme from methanotrophic bacteria has yet to be purified and its activity biochemically validated, it has been observed that in some methanotrophs ammonium can induce the expression of the genes *hao*AB (Campbell *et al.*, 2011).

The representation of methanotroph-related biomarkers was supported by 16S gene amplicon sequencing results, which revealed that methanotrophic families Methylococcaceae, Methylocystaceae and Beijerinckiaceae were almost two times higher in Ushuaia Bay than in Potter Cove (0.01%7 ± 0.013% and 0.009% ± 0.005%, respectively). Particularly in the OR site, the oil jetty, methanotrophic families were higher in abundance (0.022% ± 0.010%) This was accompanied by a higher representation of methanogenesis when analysing both community structure and metabolic potential. Families of methanogenic Archaea (Methanobacteriaceae, Methanoregulaceae and Methanosarcinaceae) were more abundant in Ushuaia Bay, especially at the OR site, where they had an abundance of 0.008% ± 0.004% (vs. non-detected in Potter Cove). Similarly, metagenomic sequences assigned to methane metabolism (map00680) had an abundance of 2.22% ± 0.33% in Ushuaia Bay and 1.63% ± 0.04% in Potter Cove (Espínola *et al.*, 2018). These results are consistent with higher loads of organic matter in Ushuaia Bay because of anthropogenic pollution in this environment, which could lead to a decrease in oxygen levels, thus favouring methanogenesis and other anaerobic processes such as sulfate reduction and anaerobic hydrocarbon biodegradation (see previous sections). The consequent production of methane gas and its migration to the upper sediment layers and water column could lead to its consumption by aerobic methanotrophic bacteria, completing its cycle. However, this process is largely affected by environmental as well as seasonal factors (Steinle *et al.*, 2017).

The comprehension of N-cycling potential in microbial communities of coastal sediments from high-latitude environments still represents a challenge because of the complexity of the microbial communities inhabiting these regions. The metagenomic approach used for these analyses was useful to reveal the potential and structure of the microbial guild associated with nitrification and related processes. Our results suggest that these communities present a high potential for hydroxylamine oxidation accompanied by a range of phylotypes, not only characterized by the typical Betaproteobacteria nitrifiers, but also methanotrophic bacteria, which appeared to be key carriers of this functional trait.

Analysis of gene content and shared synteny of algal polysaccharide utilization loci

In prokaryotes, many pathways are encoded by genes located in close proximity on the chromosome, and this gene clustering extends beyond operons (Touchon and Rocha, 2016). Because of the high rate of spontaneous genome rearrangements in these organisms, such as insertions, deletions, translocations, inversions or duplications, gene content and gene order usually have a lower level of conservation than protein sequences (Lacroix *et al.*, 2014; Rocha, 2006). Therefore, the analysis of the degree of conservation in gene cluster architecture across genomes provides evolutionary context and offers clues regarding the potential function of hypothetical proteins using the 'guilt-by-association' approach. When targeting uncultured microorganisms, however, this analysis is often hindered because of the incomplete and fragmentary nature of assembled metagenomic datasets, in particular in highly diverse microbial communities such as those from soils or sediments (Howe *et al.*, 2014; Matos *et al.*, 2016).

In this section, sequence similarity, as well as gene content and organization, were compared in clusters related to the bacterial utilization of alginate, an important carbon source for cold sediment microbial communities. Alginate is a linear polymer of 1,4-linked β-D-mannuronate (M) and α-L-guluronate (G) containing MM, GG and MG blocks in various proportions. It is abundant in brown algae and can also be synthesized by some bacteria as a defence mechanism (Zhu and Yin, 2015). In the cell wall of brown algae, alginate forms rigid matricial gels in association with phenolic compounds (Deniaud-Bouët *et al.*, 2014), which are probably difficult to access by heterotrophic marine bacteria. The first step of the degradation of alginate is the depolymerization of this polysaccharide by the action of endolytic and exolytic enzymes, alginate lyases (ALs) and oligoalginate lyases (OALs), respectively (Zhu and Yin, 2015). Until now, seven different polysaccharide lyase families [Carbohydrate-Active EnZYme Database (CAZy); Terrapon *et al.*, 2017) include AL and/or OAL enzymes, although the genes most often found in genomes from marine alginolytic bacteria belong to the PL6, PL7 and PL17 families, often in more than one copy (Thomas *et al.*, 2012; Kabisch *et al.*, 2014). The depolymerization products are transported to the cytoplasm, where a series of enzymatic steps result in the formation of ketodeoxyphosphogluconate, which is probably funneled into the Entner–Doudoroff pathway (Kabisch *et al.*, 2014). The set of genes required for alginate assimilation has been studied in detail in some strains isolated from marine habitats (e.g. Kabisch *et al.*, 2014; Takagi *et al.*, 2016; Thomas *et al.*, 2012, 2017). Gene clusters for alginate utilization in members of the phylum Bacteroidetes are examples of canonical polysaccharide utilization loci (PULs), which code for alginate-depolymerizing enzymes, membrane proteins, cytoplasmic enzymes and transcriptional factors required for the assimilation of the polysaccharide (Grondin *et al.*, 2017). However, not all genes related to alginate degradation were found to be located within PULs, as additional AL genes were located elsewhere in the chromosome (Thomas *et al.*, 2012). Alginolytic gene clusters have also been detected in other bacterial taxa (Grondin *et al.*, 2017). These clusters are usually shorter, more variable in their gene content and organization and lack the characteristic *susD* gene that codes for glycan-binding proteins located on the cell surface (Thomas *et al.*, 2012; Grondin *et al.*, 2017). Interestingly, alginate-related clusters identified in strains of *Alteromonas macleodii* (class Gammaproteobacteria, order Alteromonadales) from two different ecotypes differed in their number of AL/OAL genes, which was reflected in their alginate utilization efficiency (Neumann *et al.*, 2015).

Because of the key role that AL and OAL enzymes play in the assimilation of alginate, genes coding for these enzymes can be used as biomarkers for estimating the alginate-degrading potential of microbial communities. The abundance of genes encoding homologues of AL and OAL enzymes belonging to different polysaccharide lyase families were estimated in the metagenomes (assembled and unassembled fractions) based on the abundance of the Pfam (protein families database) domains PF05426 (alginate lyase, identified in members of the PL5 and PL17 families), PF08787 (alginate lyase 2, mostly detected in enzymes belonging to the PL7 family) and PF14592 (chondroitinase B, PL6 family). Remarkably, the sum of the abundances of these three domains in the six metagenomes from Potter Cove sediments was 1.2–2.6 times higher than the abundance of single-copy genes (average abundance of 12 genes encoding ribosomal proteins; Matos *et al.*, 2016). These results indicate that bacterial populations with the ability to utilize alginate are abundant in the sediment microbial community, which could contain several copies of AL genes in their genome. Potter Cove contains a high biomass of large brown algae species from the order Desmarestiales, such as *Desmarestia* spp. and *Himantothallus grandifolius*, which are the main primary producers in this environment (Quartino and Boraso de Zaixso, 2008). As alginate constitutes up to 40% of brown algae dry weight, the utilization of this polysaccharide represents a relevant process in the cycling of macroalgal-derived carbon in Potter Cove. Therefore, the high alginate degradation potential estimated in Potter Cove sediments is in accordance with the high abundance of alginate as a carbon source in this environment (Matos *et al.*, 2016). The abundance of the three Pfam domains was also high in Ushuaia Bay metagenomes, varying between 0.7 and 1.8 times the abundance of single-copy genes. Similarly to other cold coastal environments, Ushuaia Bay contains a high abundance of brown algae, mainly *Macrocystis pyrifera* (Liuzzi *et al.*, 2011).

We analysed the gene content, organization and shared synteny in alginate-related gene clusters identified in the metagenomic dataset. The scaffolds were selected based on the presence of at least one AL and/or OAL homologue sequence (with a minimum length of 100 amino acids) in fragments of ≥4 kb. This analysis was possible only in Potter Cove metagenomes, as Ushuaia Bay sediment metagenomes were more fragmented. The metagenomic dataset of Potter Cove sediments contained 103 scaffolds with lengths between 4 and 137 kb, which carried in total 148 AL and OAL homologue genes (out of 1141 sequences of ≥100 amino acids identified in the assembled metagenomes; Matos *et al.*, 2016). In total, 54.4% of these scaffolds were assigned to the Bacteroidetes phylum (phylogenetic lineage as determined in the IMG-M annotation pipeline; Huntemann *et al.*, 2016), with 90% of these scaffolds assigned to the Flavobacteriia class. On the other hand, 36.9% of the scaffolds were assigned to the Proteobacteria phylum (68.4% of them to the Gammaproteobacteria class and 21.1% to the Alphaprotebacteria class), 1.9% were assigned to Verrucomicrobia and 6.8% could not be assigned at the phylum level. Although these scaffolds probably do not include all members of the alginolytic guild, these results suggest that multiple bacterial populations belonging to different phyla, including highly divergent organisms, have the ability to utilize alginate. This guild seems to be dominated by organisms belonging to the Flavobacteriia and Gammaproteobacteria classes, which were also the most abundant classes in these sediments [see previous section on community structure and Espínola *et al.* (2018) and Matos *et al.* (2016)].

PULs from scaffolds assigned to the Flavobacteriia class

Up to 40 gene organizations were identified among the 56 scaffolds assigned to the Bacteroidetes phylum, suggesting a low coverage in the set of scaffolds. Only 12 of these organizations contained PULs, while the rest included AL and/or OAL homologue genes, but no other genes previously related to alginate utilization. The longest scaffold for each gene organization was selected for further analysis, using the software Trebol (see 'Web resources'). This tool allows the comparison of gene content and order, analysing the sequence similarity using discontiguous megablast (dc-megablast). This nucleotide BLAST (Basic Local Alignment Search Tool) algorithm uses parameters as word size, reward and penalty, which are suitable for inter-species comparisons (Johnson *et al.*, 2008). Eleven scaffolds, all assigned to the Flavobacteriia class, presented some level of shared synteny (Fig. 5.5).

Most PULs were fragmented because of the short length of the majority of the scaffolds, which precludes a complete analysis of gene content and organization as performed with genomes of alginolytic bacteria (Thomas *et al.*, 2012; Kabish *et al.*, 2014). PULs contained coding sequences homologous to (1) ALs and/or OALs of the PL6, PL7 and PL17 families,

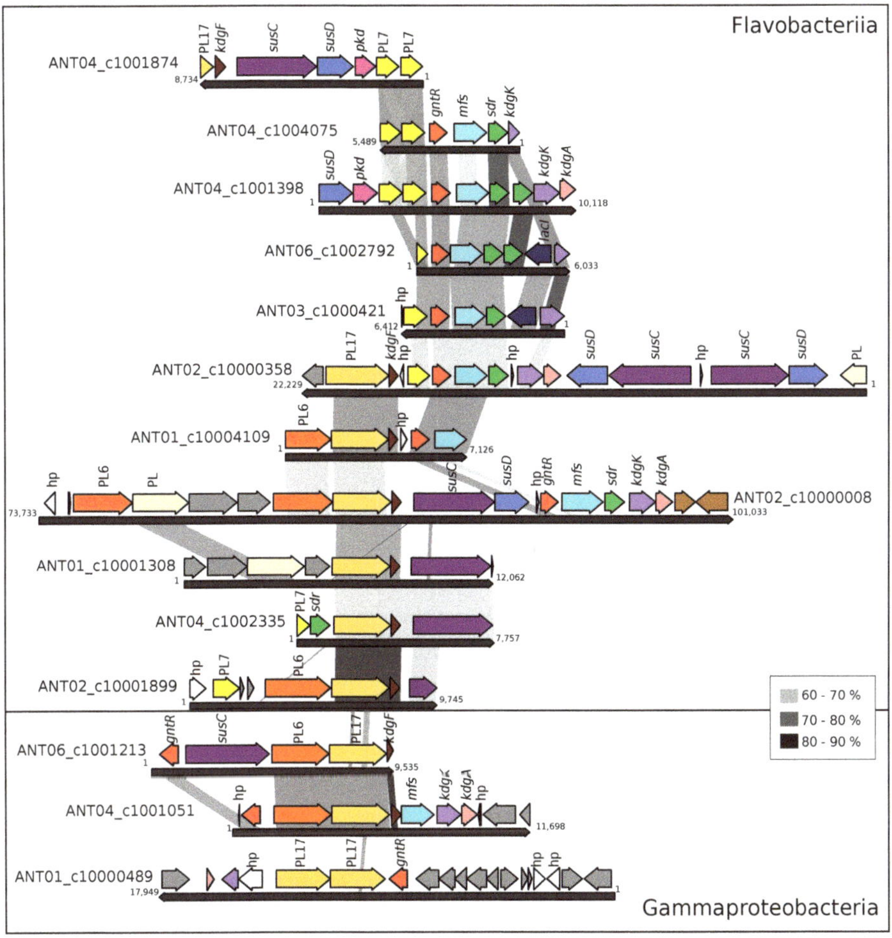

Figure 5.5 Gene organization, shared synteny and sequence similarity in alginolytic gene clusters identified in scaffolds from the Potter Cove metagenomes. For each gene cluster, the name of the sample (ANT01-04, ANT06) and the scaffold number is indicated. The dark grey arrow below each scaffold indicates the direction of the sequence, as some were inverted in order to match the orientation of the genes. The numbers at the end of the scaffolds indicate the number of bases (only a fraction of scaffold ANT02_10000008 is shown in the figure). Colours indicate different genes, with white corresponding to hypothetical proteins and grey to genes not related to polysaccharide metabolism. Ranges of identity values at the nucleotide level are shown as different shades of grey connecting scaffolds, as determined using dc-megablast in Trebol (http://bosque.udec.cl/software/trebol.html).

(2) outer membrane proteins that participate in the detection and transport of oligosaccharides across the outer membrane (SusC, SusD and PKD), (3) a permease of the cytoplasmic membrane (MFS), (4) cytoplasmic enzymes (Sdr, KdgK, KdgF and KdgA) and (5) transcriptional regulators GntR and LacI (Grondin *et al.*, 2017). These results are in agreement with the gene content of alginate-related PULs from marine alginolytic bacteria (Kabisch *et al.*, 2014; Thomas *et al.*, 2012). Hypothetical proteins mostly corresponded to short coding sequences, probably the result of an incorrect gene prediction (Fig. 5.5, white arrows). The first sequence of scaffold ANT02_c100001899 was annotated as a hypothetical protein in the IMG-M pipeline, although further analysis suggests that it could belong to the PL6 family.

PULs identified in the scaffolds assigned to the Flavobacteriia class showed a partial conservation in gene order, and syntenic fragments showed high sequence similarity values (65–87%). However, differences in gene content and order, such as gene duplications and rearrangements, were often detected. These results suggest the presence of actively evolving bacterial populations belonging to the Flavobacteriia class with the potential to assimilate alginate in Potter Cove sediments. Few genes not related to alginate assimilation were identified in these fragments, with the exception of the longer scaffold ANT02_c10000008 (Fig. 5.5, grey arrows). When comparing the gene organization of these PULs with that of PULs from the closest alginolytic strains, some level of shared synteny was observed with members of the Flavobacteriia class, although the level of conservation upstream and downstream of the PULs was scarce (data not shown). The first five scaffolds of Fig. 5.5 present a gene organization highly similar to at least part of the alginate utilization cluster of *Gramella forsetii* KT0803 (Kabish *et al.*, 2014), with the exception of the position of the *lacI* gene within the cluster and the presence of two contiguous short-chain dehydrogenase reductase (SDR) paralogues in two of the scaffolds. Based on the analysis of the genome of alginolytic bacteria isolated from different geographical locations and habitats, it has been proposed that the variability in gene number within PULs in members of the Flavobacteriia could be the result of gene duplication events, followed by the loss of one of the paralogues (Thomas *et al.*, 2102). In Potter Cove sediments, organisms with one or two copies of the SDR gene seem to coexist within the alginolytic guild. Less common among Flavobacteria is the presence of duplicated susC- and susD-like genes (Thomas *et al.*, 2012), such as scaffold ANT02_c10000358, where two sets of *susC/susD* paralogues were found to be located in opposite orientations.

Clusters from scaffolds assigned to the Gammaproteobacteria class

The 26 scaffolds assigned to Gammaproteobacteria presented 14 gene organizations, although only three of them (all assigned to the order Alteromonadales) contained PUL-like clusters (Fig. 5.5). Each of these scaffolds contained two AL/OAL homologue genes, belonging to the PL6 or PL17 families. In addition, the clusters included a *gntR*-like gene, which was located in a different orientation from the genes encoding alginate-depolymerizing enzymes, unlike clusters from scaffolds assigned to Flavobacteriia. In addition, *kdgF, kdgK, kdgA, susC* and *mfs* were identified in at least one cluster. Scaffolds ANT06_c1001213 and ANT04_c1001051 presented similar gene content and segments that shared high identity values, but the third scaffold (ANT01_c10000489) was very distinct. The shared synteny with clusters from the closest genomes, from *Psychromonas* spp. and *Colwellia* spp., was also variable (data not shown). These results are in agreement with the limited synteny previously reported in alginolytic gene clusters in members of the Proteobacteria phylum (Thomas *et al.*, 2012; Neumann *et al.*, 2015).

Alginate-related clusters from scaffolds assigned to the Gammaproteobacteria class shared low identity values with PULs from Flavobacteriia (Fig. 5.5). Interestingly, gene order was partially similar for AL and/or OAL-like genes (PL6 and PL17 families), as well as the *kdgF* gene, in some of the scaffolds from both classes. It has been suggested that alginolytic gene clusters originated in a marine flavobacterium and that one or multiple horizontal gene transfer events occurred from members of this class to other bacterial groups, such as Alphaproteobacteria and Gammaproteobacteria (Thomas *et al.*, 2012). The fact that gene organization was partially maintained, even with low shared identity values, suggests that gene order can have a relevant role in the alginolytic process, for instance for gene expression regulation.

Microbial rhodopsins of non-viral and viral origin

Microbial rhodopsins represent a wide protein family broadly distributed in microorganisms that share their basic structure with the so-called 'animal rhodopsins'. This includes an apoprotein, composed of a single polypeptide chain with seven transmembrane domains, and a cofactor consisting of a retinal molecule (the vitamin A aldehyde) that acts as a very simple light sensor. Microbial rhodopsins are also called type-I rhodopsins, to differentiate them from type-II rhodopsins from the retina, pineal gland receptor proteins, the hypothalamus and other tissues of metazoa. Besides sharing their cofactor and architecture, both rhodopsin types exhibit a very similar mechanism of action, although it is not clear whether both families have a common evolutionary origin or rather represent an example of convergent evolution. The amino acid sequence of both types of rhodopsins is highly conserved in some key short regions involved in retinal chromatophore binding (85–95% identity), although the complete sequence identity can be as low as 26% (Spudich *et al.*, 2000).

The photochemical reactions resulting from the photoisomerization of the retinal molecule drive many biological processes in different microbial rhodopsins. Biological functions can be grouped into (1) photoenergy transducers ('light-driven pumps') that convert light into an electrochemical potential that activates ion transport and (2) photosensory receptors ('sensory rhodopsins') that use ambient light to produce intracellular signals (Govorunova *et al.*, 2017). Among ion-pump rhodopsins, the most commonly found are proton pumps, for example bacteriorhodopsins (Inoue *et al.*, 2016) and proteorhodopsins (Slamovits *et al.*, 2011). Chloride and more recently sodium pumps have also been described (Kim *et al.*, 2016; Tsunoda *et al.*, 2017). On the other hand, sensory rhodopsins activate a variety of cellular signalling chains, resembling the two-component system of bacterial chemotaxis (Spudich, 2006).

Currently, more than 7000 members of this family have been described in genomes from prokaryotes, lower photosynthetic eukaryotes and protist viruses (Govorunova *et al.*, 2017). Microorganisms harbouring rhodopsin-coding genes have a global distribution, colonizing very different habitats (Gorriti *et al.*, 2014; Govorunova *et al.*, 2017; López *et al.*, 2017; Philosof and Béjà, 2013; Sabehi *et al.*, 2003; Yutin and Koonin, 2012). Proteorhodopsin homologues were described in the genome of the giant DNA viruses infecting photosynthetic protists (Organic Lake phycodnaviruses and *Phaeocystis globosa* virus) (Yutin and Koonin, 2012). To date, there is no information about the possible biological role of these viral rhodopsin genes in the hosts. Notably, the screening of sequences from metagenomes showed that viral proteorhodopsins are extremely abundant in marine environments (Philosof and Béjà, 2013).

Putative genes coding for microbial rhodopsins in cold coastal sediment metagenomes

In a recent work, a rich collection of microbial (including viral) rhodopsin genes was detected in metagenomes from high-latitude coastal sediments, including the ones analysed throughout this chapter (Potter Cove and Ushuaia Bay), as well as similar coastal environments from the northern hemisphere (López *et al.*, 2017). In this study, sequences were retrieved from the metagenomes using Clusters of Orthologous Groups (COG) (Tatusov *et al.*, 2000) as functional evidence, by searching for COG5524 ('Bacteriorhodopsin'). An initial set of 6224 sequences was obtained from this dataset. After dereplication using BlastClust, further confirmation by detection of transmembrane helices in proteins using the TMHMM server (Krogh *et al.*, 2001) and selection of sequences longer than 50 amino acids, a final set of 123 putative rhodopsin sequences was obtained, which included both viral and non-viral, sensory and light-driven proton pump proteorhodopsins. These sequences showed a wide phylogenetic origin, being associated with rhodopsins described in different bacterial phyla (Bacteroidetes, Planctomycetes, Firmicutes, Actinobacteria, Cyanobacteria, Proteobacteria and Deinococcus–Thermus), as well as microalgae (Cryptophyta), fungi and viruses (Phycodnaviridae) (López *et al.*, 2017).

A question arising from these results is what could be the ecological significance of the presence of microbial rhodopsin genes from wide phylogenetic origins in high-latitude environments where exposure to light is limited. For instance, in Potter Cove, samples were obtained from the first 5 cm of sediment, at water column depths (bathymetry) ranging from 9.5 to 50 m. This region is subjected to extreme photoperiods (large periods of continuous darkness alternating with others of continuous light). It is also characterized by heavy mixing produced by strong winds and the movement of icebergs detached from glaciers, factors limiting the stratification process of the microbial communities in the water column during summer (López *et al.*, 2017). Because of these environmental conditions, the microorganisms whose genomes harbour the rhodopsin genes do not necessarily carry out their complete biological cycle in the upper layer of the marine sediment from where they were collected, but live part of their cycle in the water column. The stratification of the water column occurring during winter, when the freezing of surface waters occurs, leads to the sedimentation of pelagic microorganisms, which could explain the presence of these genes in Potter Cove sediments. The comparison of the microbial community structures of these coastal sediments and their corresponding overlying water column by 16S rRNA gene analyses showed evidences of overlapping, supporting this hypothesis (data not shown).

Phylogeography of viral and non-viral rhodopsins

Phylogeography is the study of evolutionary processes in space and time, utilizing both the geographical and the phylogenetic history to explain current geographical patterns of organisms (Fontaneto, 2011). A statistical analysis of association between the Bayesian phylogeny and the geographical origin of the viral and non-viral microbial rhodopsin genes was carried out, with viral rhodopsin sequences from these coastal sediments together with viral GOS rhodopsin sequences (Yutin and Koonin, 2012). In order to select only sequences with a strong phylogenetic signal, 100 amino acids were used as cut-off in this case. Phylogenetic relationships among rhodopsin-coding amino acid sequences were analysed using Bayesian inference (BI) analyses, using MrBayes 3.1 software (Ronquist and Huelsenbeck, 2003). A Bayesian Tip-association Significance (BaTs) test (Parker *et al.*, 2008) was carried out based on the results of the BI. A number of statistics (association index and parsimony score, which test the global association between a trait and tree topology, as well as monophyletic clade size statistic) were calculated based on the posterior samples of trees produced by MrBayes 3.1 using the BaTs program ($n = 30{,}000$). The null distribution for each statistic was estimated with 1000 replicates of state randomization. This analysis did not evidence any phylogeographical segregation among microbial non-viral rhodopsins. In contrast, viral rhodopsin genes showed a phylogeographical segregation (López *et al.*, 2017). It is likely that non-viral genes have passed through a long period of dissemination, with numerous horizontal transfer events, determining their present global distribution without evidence of endemism. On the other hand, the finding of phylogeography in viral genes could be an indication that the acquisition of rhodopsin genes by the genome of Phycodnaviruses might be a relatively recent event with respect to the acquisition of rhodopsins by other microorganisms (bacteria, archaea, yeasts, fungi). The shorter time elapsed between the acquisition of rhodopsin genes by viruses and the present would have prevented the global dissemination of these genes, resulting in an endemic distribution. This hypothesis must be challenged in the future through coalescence studies. For this purpose, samples need to be being taken at different times during a period that, although evolutionarily short, is long enough for estimating the divergence time and age of the closest common ancestor of the present viral rhodopsin genes.

Viral rhodopsins of both group I and group II (Yutin and Koonin, 2012) were observed in the set of sequences from Potter Cove metagenomes (Fig. 5.6), clearly grouped by geographical location. In Fig. 5.6A, a distinct clade can be observed, defined with 100% of posterior probability. This clade, called PAVR (putative ancestor of viral rhodopsins), contain sequences of rhodopsin genes from Potter Cove and from other coastal environments from the Northern Hemisphere. This analysis allows is to shed light on the relationship with the probable ancestral genes of the viral rhodopsins that currently circulate in the genomes of Phycodnaviruses. These probable ancestral sequences of the current viral rhodopsins represent a strongly supported brother group containing sequences that are homologous to sequences from *Chloroflexi* and Actinobacteria (López *et al.*, 2017).

We further analysed the conservation patterns of amino acids that are key for the function of microbial rhodopsins, in the search for correlations that would allow us to better define the characteristics of the PAVR group with respect to the other two clusters, and to analyse the eventual ancestrality of the PAVR group of sequences as well as the geographical segregation observed through the phylogeographical analysis. We compared the residues around the proton donor/acceptor sites, as well as the position affecting the fine-tuning of light absorption (wavelength). Alignments were constructed with the MUSCLE online server (Edgar, 2004). Consensus sequences from each studied location were obtained using the WebLogo 3.4 online web application (Crooks *et al.*, 2004). It could be observed from this analysis that the tuning site was highly conserved, always containing a methionine residue. The proton donor site exhibited an aspartic acid residue in all analysed rhodopsins, but not in members of the PAVR group, which had a glutamic acid residue instead. Finally, the proton acceptor site was the least well conserved, with each phylogeographical group

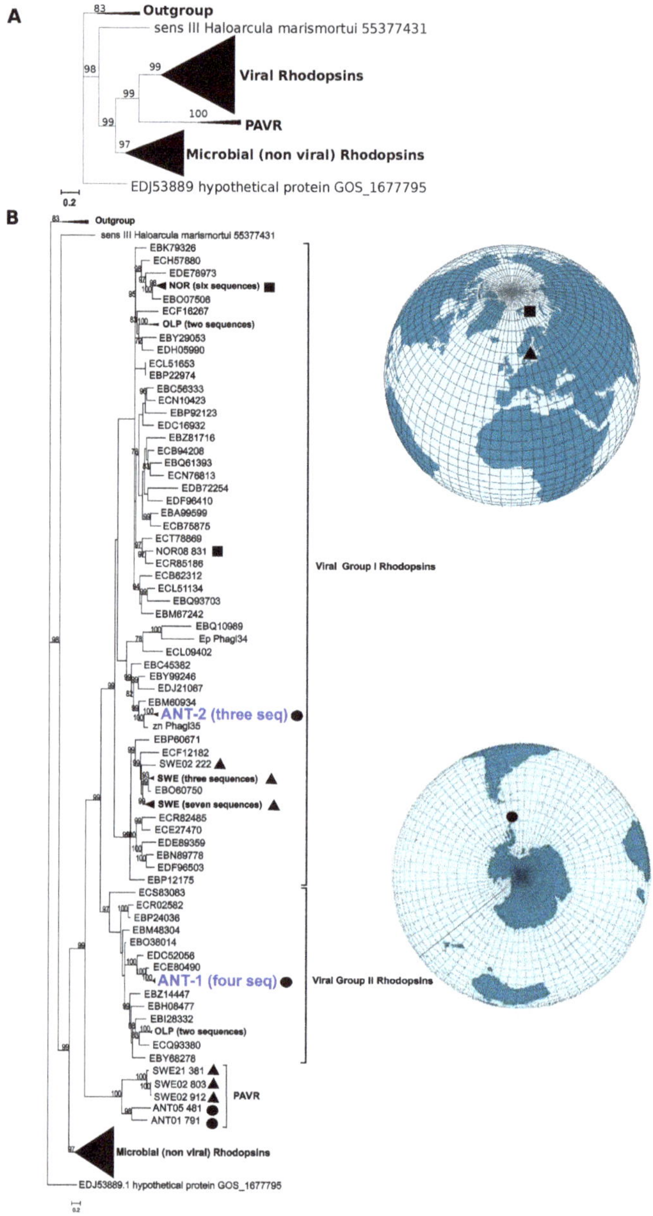

Figure 5.6 Bayesian phylogenetic reconstruction of viral rhodopsin protein sequences from Potter Cove and other polar and subpolar regions of the world. (A) Phylogenetic relationships among the main groups of microbial rhodopsins (described in López *et al.*, 2017). Note the high support values evidenced by the posterior probability inferred from the Bayesian method that support the monophyly of the viral rhodopsins, the microbial rhodopsins and the probable group of ancestral microbial sequences of the viral rhodopsins (PAVR). (B) Bayesian phylogeny of viral rhodopsins showing groups I and II, as well as viral rhodopsin sequences from high-latitude coastal sediments. ANT, Potter Cove; NOR, Adventfjord, Svalbard Archipielago, Norway; SWE, Port Värtahamnen, Stockholm, Baltic Sea, Sweden. Note the high posterior probabilities that support the phylogeographical segregation of viral rhodopsins. Viral rhodopsin protein sequences from Ushuaia Bay were not included because they did not reach the minimum length for the analysis (100 amino acids).

of viral rhodopsins containing a different amino acid in each geographical region (López *et al.*, 2017). This analysis provided additional evidence supporting both the geographical segregation of the viral sequences and the eventual ancestrality of the PAVR group.

As the structure and function of viral rhodopsins is still unknown, and given the importance of viruses in marine ecosystems (Mojica and Brussaard, 2014), it is important to increase our fundamental understanding of the role of viral photoreceptors in ecosystem functioning. In addition to shedding light on ecological and evolutionary processes affecting microbial communities, microbial rhodopsins have attracted attention because of their photoreception capacity, which make them interesting targets for heterologous expression, in particular those from viral origin. Biotechnological uses of microbial rhodopsins include, among others, the production of biofuels and biosensors, use in optogenetics and use in personal identification cards (Bacchus and Fussenegger, 2012; Gaub *et al.*, 2015; Kim *et al.*, 2012; Lee *et al.*, 2015; Minic *et al.*, 2005).

A workflow for rational selection of metagenomic sequences for heterologous expression

As shown in the previous sections, amplicon sequencing and metagenomic datasets can provide in-depth information on ecological aspects of the structure and metabolic potential of complex microbial communities. In addition to their ability to produce ecological signals, these communities contain a wealth of biotechnological potential that can be mined for their utilization in innovative processes and products. This section describes a methodological procedure that can be used to identify sequences coding for potentially useful biocatalysts from large metagenomic datasets (Musumeci *et al.*, 2017). This procedure involves the use of different bioinformatic approaches to provide evidence that allows the rational selection of sequences of putative enzymes prior to their heterologous expression and biochemical characterization. This strategy reduces the time and resources needed for the selection of biocatalysts with the desired features, by reducing the number of enzymes that need to be expressed (Ferrer *et al.*, 2016).

The selection of sequences coding for Baeyer–Villiger monooxygenases (BVMOs) from the previously mentioned metagenomic dataset will be used for illustrative purposes. BVMOs catalyse the enantioselective oxidation of ketones to produce esters or lactones and they have been applied in the production of chiral intermediates for the synthesis of natural products and analogues (Alphand and Wohlgemuth, 2010; Ceccoli *et al.*, 2014). These enzymes belong to the group B flavin-dependent monooxygenases (B-FDMs), a family of enzymes that also includes flavoprotein monooxygenases (FMOs), N-hydroxylating monooxygenases (NHMOs) and YUCCA monooxygenases (indole 3-pyruvate monooxygenase, EC 1.14.13.168) (Huijbers *et al.*, 2014).

The first step of the procedure involves the identification of conserved domains contained in the prospected biocatalysts (Fig. 5.7, step 1). The Conserved Domain Database (CDD) in NCBI can be searched to identify the specific domains in reference sequences, for instance those encoding enzymes that have been previously characterized. This can be performed using either the *CD-Search* or the *Batch CD-Search* tool. This information is also displayed when performing the alignment of a sequence against the NCBI non redundant database using BLAST (Altschul *et al.*, 1997). A list of domain hits with the corresponding accession codes for the protein family, such as COG (Tatusov *et al.*, 2000), Pfam (Finn *et al.*, 2016)

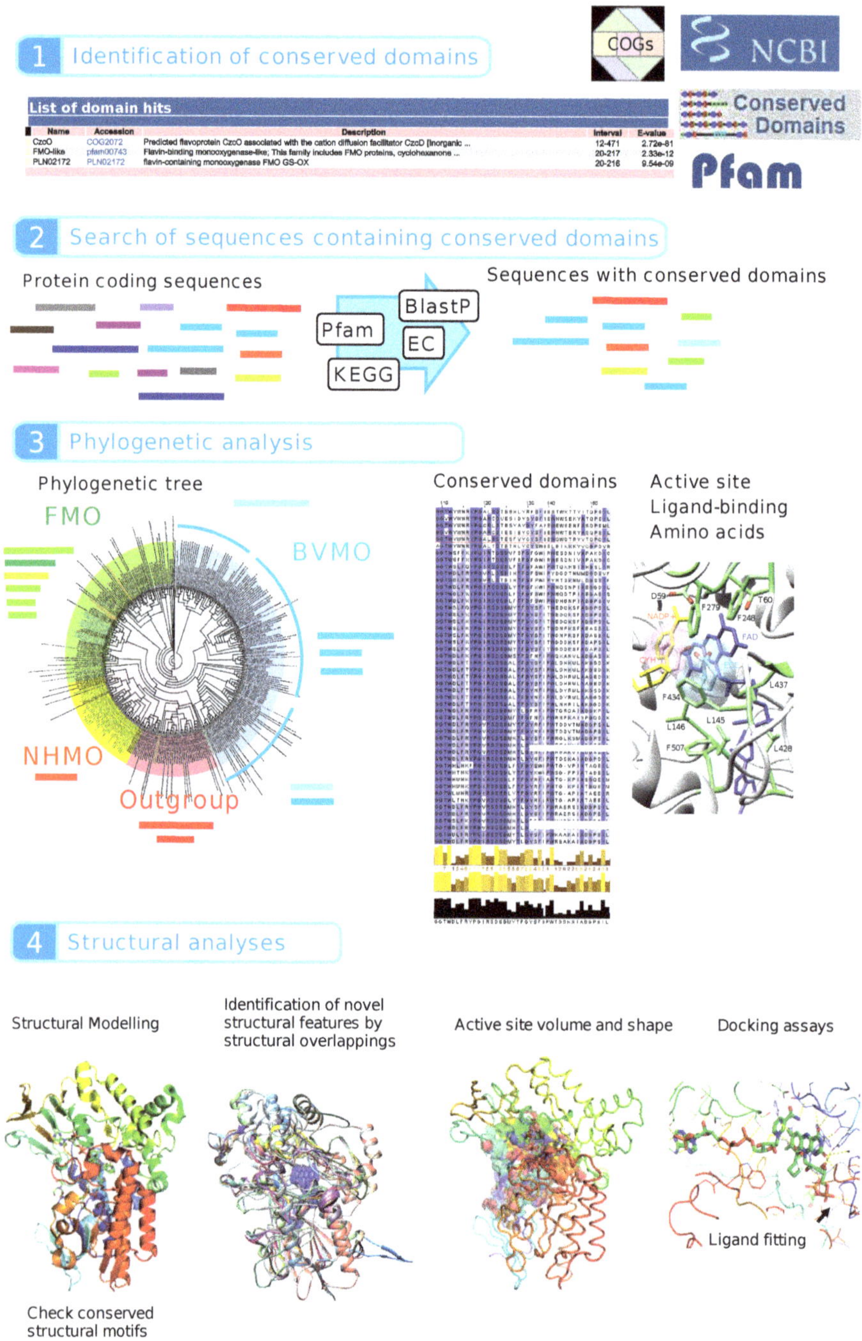

Figure 5.7 Overview of a methodological procedure for the selection of metagenomic sequences coding for enzymes with potential novel features, for biotechnological applications. Steps 1–4 are the sequential steps of the workflow.

and TIGRFAMs, is obtained (Fig. 5.7). For example, the Pfam domain PF00743 (flavin-dependent monooxygenase like) is present, although not exclusively, in all BVMO enzymes and therefore constitutes functional evidence that can be used to perform the search. The metagenomic dataset is then mined for sequences encoding the desired enzymes (Fig. 5.7, step 2). Based on the specificity of the conserved domain, searches may need to be further refined. Other potential approaches include (1) the use of BLAST against the metagenomic sequences as database, with reference sequence(s) as query; (2) the construction of specific hidden Markov models from an alignment of reference sequences using *hmmbuild* and its use for mining the dataset using the *hmmsearch* tool in HMMER3 (Eddy, 2009) or (3) the annotation of the metagenomic sequences through ghostKOALA to obtain KEGG Orthologues and EC numbers (Kanehisa *et al.*, 2016). The goal of the initial mining is to obtain the largest possible number of sequences, even those from distant homologues, which will later be analysed in more detail.

In this work, the search was performed using the Pfam domain PF0074, as well as BLAST (cut-off E-value of 10^{-5}) with sequences of well-characterized BVMOs as queries against the metagenomic dataset deposited at IMG, followed by the selection of sequences longer than 250 amino acids (Musumeci *et al.*, 2017). These methods identified seven metagenomic sequences in the six metagenomes from Ushuaia Bay and 53 sequences in the metagenomes of Potter Cove, potentially coding for BVMOs, out of 5.55×10^6 protein-coding sequences in the assembled fraction of these metagenomes. Some of the identified sequences, however, could correspond to other enzymes from the same family containing this Pfam domain. In order to select putative BVMOs out of this first selection pool, a phylogenetic analysis was carried out, including the metagenomic sequences, their closest BLASTP matches against the Representative Genomes database and well-established reference sequences belonging to B-FDM. Briefly, sequences were aligned in ClustalX, manually trimmed and a maximum-likelihood phylogenetic tree was constructed in RAxML version 8.2.3 using the GAMMA model of rate heterogeneity, LG substitution matrix and empirical base frequencies (option PROTGAMMALGF). Bootstrapping was performed with 100 replications (Musumeci *et al.*, 2017). The classification of metagenomic sequences as BVMO, FMO or NHMO was performed based on their affiliation with well-supported clusters (> 50% bootstrap) formed by the different subfamilies of B-FDM, with respect to an outgroup formed by class A FDMs (Fig. 5.7, step 3). The presence of conserved residues characteristic of these enzymes, such as those involved in the catalytic mechanism, was verified in order to confirm the classification. For example, BVMO enzymes are characterized by the presence of two Rossmann motifs (GxGxx[G/A]) flanking two BVMO fingerprints ([A/G]GxWxxxx[F/Y]P[G/M]xxxD and FxGxxxHxxxW[P/D]) (Fraaije *et al.*, 2002; Mascotti *et al.*, 2015; Riebel *et al.*, 2012). The presence of these residues was assessed in both the sequence alignment (primary structure) and in the three-dimensional (3-D) model of the sequence (tertiary structure, detailed below) (Fig. 5.7, step 3). After this step, only seven sequences remained out of the 60 metagenomic sequences originally identified, which were selected for additional analyses.

The fourth step involved a detailed analysis of the modelled 3-D structures of selected metagenomic sequences. Structural analyses provide information that can be relevant for the selection of biocatalysts, such as potential stability, the presence of prosthetic groups and probable substrate specificity. Homology modelling (also known as comparative modelling or template-based modelling) refers to 3-D structure prediction of a query sequence by its comparison with an evolutionary related (homologous) template. It is based on the concept that, during evolution, the 3-D structure is more stable than the primary structure (amino acid sequence). This implies that even distantly related sequences fold into similar structures, provided they fall in the 'safe' prediction zone (length and per cent identity threshold with respect to template) (Krieger *et al.*, 2005). This approach has the advantage that it can be carried out in an automated manner using servers such as the Swiss Model (Biasini *et al.*, 2014) or Phyre2 (Kelley *et al.*, 2015) that are able to process several sequences simultaneously. Structural models can also be built using standalone software, such as Modeller (Webb and Sali, 2014). In this case, sequence alignments with template structures and others parameters can be adjusted to increase the quality of the 3-D models. In homology modelling, the selection of the template is key for obtaining structural models with acceptable quality. The MAFFT server (Katoh *et al.*, 2017) can be used for obtaining a list of potential templates based on resolution, coverage and similarity with the query sequence. The quality of the models needs to be assessed based on their stereochemical properties, for example using RAMPAGE or PROCHEK (Laskowski *et al.*, 1993).

Informative analyses of the 3-D structure include characteristics of their catalytic pocket, which can be obtained using the software CAVER (Kozlikova *et al.*, 2014) or the server CASTp (Dundas *et al.*, 2006). These tools provide information about size, shape and volume of the catalytic pocket, as well as insights on substrate accessibility through the identification of cavities tunneling to catalytic pockets. The overlap of a metagenomic 3-D model with a known structure such as the template used for constructing the model (e.g. in UCSF Chimera; Pettersen *et al.*, 2004) allows the identification of residues potentially relevant for catalysis, substrate interaction or stereo-selectivity. Residues essential for catalysis usually show a high degree of conservation. However, if differences are observed in residues involved in substrate interaction, differences in substrate profile and/or stereo-selectivity could be expected. Novel features in the modelled metagenomic sequences could also be identified by structural comparison, such as the presence of additional domains, hairpins or novel structural loops. On the other hand, molecular-docking analyses (e.g. AutoDock Vina; Trott and Olson, 2010) can be used to predict the potential substrate specificity of the modelled metagenomic enzymes. The main limitation in these studies is the limited availability of structures of enzymes interacting with the substrate molecule, which provides a supported reference of productive binding. Although docking analysis can be performed without a previously crystallized enzyme–ligand structure, reference protein–ligand complexes provide more precision to the predicted substrate poses.

Using the described procedures, it was observed that two of the three residues relevant for catalysis were fully conserved in the seven metagenomic structural models (Musumeci *et al.*, 2017). However, tyrosine was found instead of the conserved catalytic tryptophan involved in NADPH interaction (W492) in six of the seven metagenomic sequences, a difference that preserved the aromatic nature of the key residue. Among them, two sequences showed additional structural features that resulted in important differences from the template structures, including a higher accessibility to the catalytic site, larger binding pockets and wider substrate range according to docking analysis (Musumeci *et al.*, 2017). The rational workflow described in this work allowed the selection of two sequences with promising features out of 60 originally identified in a large albeit fragmented metagenomic shotgun dataset. In addition to aiding in the selection of sequences for heterologous expression and characterization, the results obtained from these analyses can guide the further engineering of these enzymes to obtain a particular chemical product. For example, the active site of the

BVMO from *Arthrobacter* sp. BP2 was redesigned though mutational and docking analyses, completely switching the regioselectivity of the enzyme to match the desired function of the pharmaceutical industry (Balke *et al.*, 2016).

Conclusions

Microbial communities are a source of valuable environmental information that can be used to increase our understanding of the ecosystem processes that occur in natural environments. In this chapter, we presented a case study involving the metagenomic analysis of high-latitude coastal environments from the southernmost regions of the world, focusing on the role of sediment communities in carbon and nitrogen cycles, as well as other environmentally relevant processes. Analysis of the structure and metabolic potential of these microbial communities enabled us to shed light on the identity and capabilities of the different microbial populations possibly participating in ecosystem services. Multiple-level analyses, especially gene context and pathway-level information, offered mechanistic and evolutionary insights into these processes. The use of a bioinformatic/statistical toolbox originally borrowed from genomics and community ecology, and further adapted to the particular field of metagenomics, were key for the access of this wealth of information. A wide array of software resources and public databases was presented, yet numerous more are available and are continually being developed for this purpose. In addition, these tools allowed us to explore the biotechnological potential contained in this rich genetic reservoir.

Web resources

All websites last accessed 28 November 2018.

- Archaeopteryx (https://sites.google.com/site/cmzmasek/home/software/archaeopteryx-js)
- AutoDock Vina (http://vina.scripps.edu/)
- CASTP (http://sts.bioe.uic.edu/castp/)
- CAVER (http://www.caver.cz/)
- ClustalX (http://www.clustal.org/clustal2/)
- Cluster of Orthologous Groups (https://www.ncbi.nlm.nih.gov/COG/)
- Conserved Domain Database (CDD) (https://www.ncbi.nlm.nih.gov/Structure/cdd/cdd.shtml)
- Earth Microbiome Project (www.earthmicrobiome.org/)
- GhostKOALA (https://www.kegg.jp/ghostkoala/)
- Greengenes Database (http://greengenes.secondgenome.com/)
- Guppy (https://matsen.github.io/pplacer/generated_rst/guppy.html)
- HMMER3 (https://www.ebi.ac.uk/Tools/hmmer/)
- Integrated Microbial Genomes with Microbiome Samples (IMG/M) (https://img.jgi.doe.gov/m/)
- KEGG Orthology (KO) Terms and Pathways (http://www.genome.jp/kegg/kegg2.html)
- MAFFT (https://mafft.cbrc.jp/alignment/server/)
- Modeller (https://salilab.org/modeller/)
- PFAM database (http://pfam.xfam.org/)

- Phyre2 (http://www.sbg.bio.ic.ac.uk/~)
- PROCHECK (https://www.ebi.ac.uk/)
- QIIME (https://qiime2.org)
- RAMPAGE (http://mordred.bioc.cam.ac.uk/)
- RaxML (https://github.com/stamatak/standard-RAxML)
- Representative Genomes Database (https://www.ncbi.nlm.nih.gov/genome/browse/representative/)
- Standalone BLAST (https://www.ncbi.nlm.nih.gov/books/NBK279690/)
- SwissModel (https://swissmodel.expasy.org/)
- TMHMM Server v. 2.0 (http://www.cbs.dtu.dk/services/TMHMM/)
- TIGRFAMs (http://www.jcvi.org/cgi-bin/tigrfams/index.cgi)
- Trebol (http://bosque.udec.cl/software/trebol.html)
- UCSF (Chimera https://www.cgl.ucsf.edu/)
- WebLogo (http://weblogo.berkeley.edu/)

Acknowledgements

ML, HMD and MM are staff members from the Argentinean National Research Council (CONICET). PC, FE and JG are fellows from CONICET. The amplicon dataset was generated by the Earth Microbiome Project (www.earthmicrobiome.org/). The metagenomic dataset was generated at the Department of Energy–Joint Genome Institute (DOE-JGI) under the Community Sequencing Program (CSP proposal ID 328, project IDs 403959, 404206, 404777, 404782, 404786, 404788, 404801). This research was also partly funded by CONICET (project no. 112–200801–01736) and ANPCyT (PICT 2016–2101, PICT 2015–2102 and PICTO 2010–0124), by the Universidad de Buenos Aires (project 20020130100569BA) and under the Laboratory Directed Research and Development Program at PNNL, a multi-program national laboratory operated by Battelle for the US Department of Energy under Contract DE-AC05–76RL01830.

References

Abele, D., Vazquez, S., Buma, A.G.J., Hernandez, E., Quiroga, C., Held, C., Frickenhaus, S., Harms, L., Lopez, J.L., Helmke, E., *et al.* (2017). Pelagic and benthic communities of the Antarctic ecosystem of Potter Cove: Genomics and ecological implications. Mar. Genomics 33, 1–11.

Acosta-González, A., Rosselló-Móra, R., and Marqués, S. (2013). Diversity of benzylsuccinate synthase-like (bssA) genes in hydrocarbon-polluted marine sediments suggests substrate-dependent clustering. Appl. Environ. Microbiol. *79*, 3667–3676. https://doi.org/10.1128/AEM.03934-12

Alphand, V., and Wohlgemuth, R. (2010). Applications of Baeyer-Villiger monooxygenases in organic synthesis. Curr. Org. Chem. *14*, 1928–1965.

Altschul, S.F., Madden, T.L., Schäffer, A.A., Zhang, J., Zhang, Z., Miller, W., and Lipman, D.J. (1997). Gapped BLAST and PSI-BLAST: a new generation of protein database search programs. Nucleic Acids Res. *25*, 3389–3402.

Bacchus, W., and Fussenegger, M. (2012). The use of light for engineered control and reprogramming of cellular functions. Curr. Opin. Biotechnol. *23*, 695–702. https://doi.org/10.1016/j.copbio.2011.12.004

Balke, K., Schmidt, S., Genz, M., and Bornscheuer, U.T. (2016). Switching the regioselectivity of a cyclohexanone monooxygenase toward (+)-*trans*-dihydrocarvone by rational protein design. ACS Chem. Biol. *11*, 38–43. https://doi.org/10.1021/acschembio.5b00723

Barbier, E.B. (2017). Marine ecosystem services. Curr. Biol. *27*, R507–R510.

Bargiela, R., Mapelli, F., Rojo, D., Chouaia, B., Tornés, J., Borin, S., Richter, M., Del Pozo, M.V., Cappello, S., Gertler, C., *et al.* (2015). Bacterial population and biodegradation potential in chronically crude oil-contaminated marine sediments are strongly linked to temperature. Sci. Rep. *5*, 11651. https://doi.org/10.1038/srep11651

Bauer, J.E., Cai, W.J., Raymond, P.A., Bianchi, T.S., Hopkinson, C.S., and Regnier, P.A. (2013). The changing carbon cycle of the coastal ocean. Nature *504*, 61–70. https://doi.org/10.1038/nature12857

Berger, S.A., Krompass, D., and Stamatakis, A. (2011). Performance, accuracy, and Web server for evolutionary placement of short sequence reads under maximum likelihood. Syst. Biol. *60*, 291–302. https://doi.org/10.1093/sysbio/syr010

Bergmann, D.J., Hooper, A.B., and Klotz, M.G. (2005). Structure and sequence conservation of *hao* cluster genes of autotrophic ammonia-oxidizing bacteria: evidence for their evolutionary history. Appl. Environ. Microbiol. *71*, 5371–5382.

Biasini, M., Bienert, S., Waterhouse, A., Arnold, K., Studer, G., Schmidt, T., Kiefer, F., Cassarino, T.G., Bertoni, M., and Bordoli, L. (2014). SWISS-MODEL: modelling protein tertiary and quaternary structure using evolutionary information. Nucleic Acids Res. *42*, W252–W258.

Biers, E.J., Sun, S., and Howard, E.C. (2009). Prokaryotic genomes and diversity in surface ocean waters: interrogating the global ocean sampling metagenome. Appl. Environ. Microbiol. *75*, 2221–2229. https://doi.org/10.1128/AEM.02118-08

Bothe, H., Jost, G., Schloter, M., Ward, B.B., and Witzel, K. (2000). Molecular analysis of ammonia oxidation and denitrification in natural environments. FEMS Microbiol. Rev. *24*, 673–690.

Bray, J.R., and Curtis, C.T. (1957). An ordination of the upland forest communities of Southern Wisconsin. Ecol. Monogr. *27*, 325–349.

Bujalesky, G.G. (2007). Coastal geomorphology and evolution of Tierra del Fuego (Southern Argentina). Geol. Acta *5*, 337–362.

Callaghan, A.V., Davidova, I.A., Savage-Ashlock, K., Parisi, V.A., Gieg, L.M., Suflita, J.M., Kukor, J.J., and Wawrik, B. (2010). Diversity of benzyl- and alkylsuccinate synthase genes in hydrocarbon-impacted environments and enrichment cultures. Environ. Sci. Technol. *44*, 7287–7294. https://doi.org/10.1021/es1002023

Calvó, L., and Garcia-Gil, L.J. (2004). Use of amoB as a new molecular marker for ammonia-oxidizing bacteria. J. Microbiol. Methods *57*, 69–78. https://doi.org/10.1016/j.mimet.2003.11.019

Campbell, M.A., Nyerges, G., Kozlowski, J.A., Poret-Peterson, A.T., Stein, L.Y., and Klotz, M.G. (2011). Model of the molecular basis for hydroxylamine oxidation and nitrous oxide production in methanotrophic bacteria. FEMS Microbiol. Lett. *322*, 82–89. https://doi.org/10.1111/j.1574-6968.2011.02340.x

Caporaso, J.G., Kuczynski, J., Stombaugh, J., Bittinger, K., Bushman, F.D., Costello, E.K., Fierer, N., Peña, A.G., Goodrich, J.K., Gordon, J.I., *et al.* (2010). QIIME allows analysis of high-throughput community sequencing data. Nat. Methods *7*, 335–336. https://doi.org/10.1038/nmeth.f.303

Caporaso, J.G., Lauber, C.L., Walters, W.A., Berg-Lyons, D., Huntley, J., Fierer, N., Owens, S.M., Betley, J., Fraser, L., Bauer, M., *et al.* (2012). Ultra-high-throughput microbial community analysis on the Illumina HiSeq and MiSeq platforms. ISME J. *6*, 1621–1624. https://doi.org/10.1038/ismej.2012.8

Cavicchioli, R. (2015). Microbial ecology of Antarctic aquatic systems. Nat. Rev. Microbiol. *13*, 691–706. https://doi.org/10.1038/nrmicro3549

Ceccoli, R.D., Bianchi, D.A., and Rial, D.V. (2014). Flavoprotein monooxygenases for oxidative biocatalysis: recombinant expression in microbial hosts and applications. Front. Microbiol. *5*, 25. https://doi.org/10.3389/fmicb.2014.00025

Commendatore, M.G., and Esteves, J.L. (2007). An assessment of oil pollution in the coastal zone of patagonia, Argentina. Environ. Manage. *40*, 814–821. https://doi.org/10.1007/s00267-005-0221-3

Commendatore, M.G., Nievas, M.L., Amin, O., and Esteves, J.L. (2012). Sources and distribution of aliphatic and polyaromatic hydrocarbons in coastal sediments from the Ushuaia Bay (Tierra del Fuego, Patagonia, Argentina). Mar. Environ. Res. *74*, 20–31. https://doi.org/10.1016/j.marenvres.2011.11.010

Cravo-Laureau, C., and Duran, R. (2014). Marine coastal sediments microbial hydrocarbon degradation processes: contribution of experimental ecology in the omics'era. Front. Microbiol. *5*, 39. https://doi.org/10.3389/fmicb.2014.00039

Crooks, G.E., Hon, G., Chandonia, J.M., and Brenner, S.E. (2004). WebLogo: a sequence logo generator. Genome Res. *14*, 1188–1190. https://doi.org/10.1101/gr.849004

Curtosi, A., Pelletier, E., Vodopivez, C., St Louis, R., and Mac Cormack, W.P. (2010). Presence and distribution of persistent toxic substances in sediments and marine organisms of Potter Cove, Antarctica. Arch. Environ. Contam. Toxicol. *59*, 582–592. https://doi.org/10.1007/s00244-010-9509-2

Daims, H., Lebedeva, E.V., Pjevac, P., Han, P., Herbold, C., Albertsen, M., Jehmlich, N., Palatinszky, M., Vierheilig, J., Bulaev, A., *et al.* (2015). Complete nitrification by Nitrospira bacteria. Nature *528*, 504–509. https://doi.org/10.1038/nature16461

Dauner, A.L., Hernández, E.A., MacCormack, W.P., and Martins, C.C. (2015). Molecular characterisation of anthropogenic sources of sedimentary organic matter from Potter Cove, King George Island, Antarctica. Sci. Total Environ. *502*, 408–416. https://doi.org/10.1016/j.scitotenv.2014.09.043

Delmont, T.O., Robe, P., Clark, I., Simonet, P., and Vogel, T.M. (2011). Metagenomic comparison of direct and indirect soil DNA extraction approaches. J. Microbiol. Methods *86*, 397–400. https://doi.org/10.1016/j.mimet.2011.06.013

Deniaud-Bouët, E., Kervarec, N., Michel, G., Tonon, T., Kloareg, B., and Hervé, C. (2014). Chemical and enzymatic fractionation of cell walls from Fucales: insights into the structure of the extracellular matrix of brown algae. Ann. Bot. *114*, 1203–1216. https://doi.org/10.1093/aob/mcu096

DeSantis, T.Z., Hugenholtz, P., Larsen, N., Rojas, M., Brodie, E.L., Keller, K., Huber, T., Dalevi, D., Hu, P., and Andersen, G.L. (2006). Greengenes, a chimera-checked 16S rRNA gene database and workbench compatible with ARB. Appl. Environ. Microbiol. *72*, 5069–5072.

Dionisi, H.M., Lozada, M., Marcos, M.S., Di Marzio, W.D., and Loviso, C.L. (2011). Aromatic hydrocarbon degradation genes from chronically polluted Subantarctic marine sediments. In Handbook of Molecular Microbial Ecology II: Metagenomics in Different Habitats, de Bruijn, F.J., ed. (John Wiley & Sons, Inc., Hoboken, NJ), pp. 461–473. https://dx.doi.org/10.1002/9781118010549.ch45

Duarte, C.A., Giarratano, E., and Gil, M.N. (2012). Trace metal content in sediments and autochthonous intertidal organisms from two adjacent bays near Ushuaia, Beagle Channel (Argentina). Mar. Environ. Res. *79*, 55–62. https://doi.org/10.1016/j.marenvres.2012.05.003

Dundas, J., Ouyang, Z., Tseng, J., Binkowski, A., Turpaz, Y., and Liang, J. (2006). CASTp: computed atlas of surface topography of proteins with structural and topographical mapping of functionally annotated residues. Nucleic Acids Res. *34*, W116–W118.

Eddy, S.R. (2009). A new generation of homology search tools based on probabilistic inference. Genome Inform. *23*, 205–211.

Edgar, R.C. (2004). MUSCLE: multiple sequence alignment with high accuracy and high throughput. Nucleic Acids Res. *32*, 1792–1797. https://doi.org/10.1093/nar/gkh340

Espínola, F., Dionisi, H.M., Borglin, S., Brislawn, C.J., Jansson, J.K., Mac Cormack, W.P., Carroll, J., Sjöling, S., and Lozada, M. (2018). Metagenomic analysis of subtidal sediments from polar and subpolar coastal environments highlights the relevance of anaerobic hydrocarbon degradation processes. Microb. Ecol. *75*, 123–139.

Ferrer, M., Martínez-Martínez, M., Bargiela, R., Streit, W.R., Golyshina, O.V., and Golyshin, P.N. (2016). Estimating the success of enzyme bioprospecting through metagenomics: current status and future trends. Microb. Biotechnol. *9*, 22–34. https://doi.org/10.1111/1751-7915.12309

Figuerola, E.L., and Erijman, L. (2010). Diversity of nitrifying bacteria in a full-scale petroleum refinery wastewater treatment plant experiencing unstable nitrification. J. Hazard. Mater. *181*, 281–288. https://doi.org/10.1016/j.jhazmat.2010.05.009

Finn, R.D., Coggill, P., Eberhardt, R.Y., Eddy, S.R., Mistry, J., Mitchell, A.L., Potter, S.C., Punta, M., Qureshi, M., Sangrador-Vegas, A., *et al.* (2016). The Pfam protein families database: towards a more sustainable future. Nucleic Acids Res. *44*, D279–85. https://doi.org/10.1093/nar/gkv1344

Fraaije, M.W., Kamerbeek, N.M., van Berkel, W.J., and Janssen, D.B. (2002). Identification of a Baeyer-Villiger monooxygenase sequence motif. FEBS Lett. *518*, 43–47.

Fontaneto, D. (2011). Biogeography of microscopic organisms: is everything small everywhere? (Cambridge University Press, New York).

Gaub, B.M., Berry, M.H., Holt, A.E., Isacoff, E.Y., and Flannery, J.G. (2015). Optogenetic Vision Restoration Using Rhodopsin for Enhanced Sensitivity. Mol. Ther. *23*, 1562–1571. https://doi.org/10.1038/mt.2015.121

Giarratano, E., Duarte, C.A., and Amin, O.A. (2010). Biomarkers and heavy metal bioaccumulation in mussels transplanted to coastal waters of the Beagle Channel. Ecotoxicol. Environ. Saf. *73*, 270–279. https://doi.org/10.1016/j.ecoenv.2009.10.009

Gieseke, A., Bjerrum, L., Wagner, M., and Amann, R. (2003). Structure and activity of multiple nitrifying bacterial populations co-existing in a biofilm. Environ. Microbiol. *5*, 355–369.

Gihring, T.M., Green, S.J., and Schadt, C.W. (2012). Massively parallel rRNA gene sequencing exacerbates the potential for biased community diversity comparisons due to variable library sizes. Environ. Microbiol. *14*, 285–290. https://doi.org/10.1111/j.1462-2920.2011.02550.x

Gil, M.N., Torres, A.I., Amin, O., and Esteves, J.L. (2011). Assessment of recent sediment influence in an urban polluted subantarctic coastal ecosystem. Beagle Channel (Southern Argentina). Mar. Pollut. Bull. *62*, 201–207. https://doi.org/10.1016/j.marpolbul.2010.10.004

Gilbert, J.A., Jansson, J.K., and Knight, R. (2014). The Earth Microbiome project: successes and aspirations. BMC Biol. *12*, 69. https://doi.org/10.1186/s12915-014-0069-1

Gorriti, M.F., Dias, G.M., Chimetto, L.A., Trindade-Silva, A.E., Silva, B.S., Mesquita, M.M., Gregoracci, G.B., Farias, M.E., Thompson, C.C., and Thompson, F.L. (2014). Genomic and phenotypic attributes of novel salinivibrios from stromatolites, sediment and water from a high altitude lake. BMC Genomics *15*, 473. https://doi.org/10.1186/1471-2164-15-473

Govorunova, E.G., Sineshchekov, O.A., Li, H., and Spudich, J.L. (2017). Microbial rhodopsins: diversity, mechanisms, and optogenetic applications. Annu. Rev. Biochem. *86*, 845–872. https://doi.org/10.1146/annurev-biochem-101910-144233

Grondin, J.M., Tamura, K., Déjean, G., Abbott, D.W., and Brumer, H. (2017). Polysaccharide utilization loci: fuelling microbial communities. J. Bacteriol. *199*, e00860-16. doi:10.1128/jb.00860-16

Groudieva, T., Grote, R., and Antranikian, G. (2003). *Psychromonas arctica* sp. nov., a novel psychrotolerant, biofilm-forming bacterium isolated from Spitzbergen. Int. J. Syst. Evol. Microbiol. *53*, 539–545.

Guibert, L.M., Loviso, C.L., Borglin, S., Jansson, J.K., Dionisi, H.M., and Lozada, M. (2016). Diverse bacterial groups contribute to the alkane degradation potential of chronically polluted subantarctic coastal sediments. Microb. Ecol. *71*, 100–112. https://doi.org/10.1007/s00248-015-0698-0

Guibert, L.M., Loviso, C.L., Marcos, M.S., Commendatore, M.G., Dionisi, H.M., and Lozada, M. (2012). Alkane biodegradation genes from chronically polluted subantarctic coastal sediments and their shifts in response to oil exposure. Microb. Ecol. *64*, 605–616. Haegeman, B., Hamelin, J., Moriarty, J., Neal, P., Dushoff, J., and Weitz, J.S. (2013). Robust estimation of microbial diversity in theory and in practice. ISME J. *7*, 1092–1101. https://doi.org/10.1038/ismej.2013.10

Hallin, S., Philippot, L., Löffler, F.E., Sanford, R.A., and Jones, C.M. (2017). Genomics and ecology of novel N_2O-reducing microorganisms. Trends Microbiol. *26*, 43–55. doi: 10.1016/j.tim.2017.07.003

Han, M.V., and Zmasek, C.M. (2009). phyloXML: XML for evolutionary biology and comparative genomics. BMC Bioinformatics *10*, 356. https://doi.org/10.1186/1471-2105-10-356

Head, I.M., Hiorns, W.D., Embley, T.M., McCarthy, A.J., and Saunders, J.R. (1993). The phylogeny of autotrophic ammonia-oxidizing bacteria as determined by analysis of 16S ribosomal RNA gene sequences. J. Gen. Microbiol. *139*, 1147–1153. https://doi.org/10.1099/00221287-139-6-1147

Head, I.M., Jones, D.M., and Röling, W.F. (2006). Marine microorganisms make a meal of oil. Nat. Rev. Microbiol. *4*, 173–182.

Herbert, R.A. (1999). Nitrogen cycling in coastal marine ecosystems. FEMS Microbiol. Rev. *23*, 563–590.

Howe, A.C., Jansson, J.K., Malfatti, S.A., Tringe, S.G., Tiedje, J.M., and Brown, C.T. (2014). Tackling soil diversity with the assembly of large, complex metagenomes. Proc. Natl. Acad. Sci. U.S.A. *111*, 4904–4909. https://doi.org/10.1073/pnas.1402564111

Huettel, M., Berg, P., and Kostka, J.E. (2014). Benthic exchange and biogeochemical cycling in permeable sediments. Ann. Rev. Mar. Sci. *6*, 23–51. https://doi.org/10.1146/annurev-marine-051413-012706

Huijbers, M.M., Montersino, S., Westphal, A.H., Tischler, D., and van Berkel, W.J. (2014). Flavin dependent monooxygenases. Arch. Biochem. Biophys. *544*, 2–17. https://doi.org/10.1016/j.abb.2013.12.005

Huntemann, M., Ivanova, N.N., Mavromatis, K., Tripp, H.J., Paez-Espino, D., Tennessen, K., Palaniappan, K., Szeto, E., Pillay, M., Chen, I.M., *et al.* (2016). The standard operating procedure of the DOE-JGI Metagenome Annotation Pipeline (MAP v.4). Stand. Genomic Sci. *11*, 17. https://doi.org/10.1186/s40793-016-0138-x

Inoue, K., Ito, S., Kato, Y., Nomura, Y., Shibata, M., Uchihashi, T., Tsunoda, S.P., and Kandori, H. (2016). A natural light-driven inward proton pump. Nat. Commun. 7, 13415. https://doi.org/10.1038/ncomms

Johnson, M., Zaretskaya, I., Raytselis, Y., Merezhuk, Y., McGinnis, S., and Madden, T.L. (2008). NCBI BLAST: a better web interface. Nucleic Acids Res. *36*, W5–W9.

Kabisch, A., Otto, A., König, S., Becher, D., Albrecht, D., Schüler, M., Teeling, H., Amann, R.I., and Schweder, T. (2014). Functional characterization of polysaccharide utilization loci in the marine Bacteroidetes '*Gramella* forsetii' KT0803. ISME J. *8*, 1492–1502. https://doi.org/10.1038/ismej.2014.4

Kanehisa, M. (2002). The KEGG database. Novartis Found. Symp. *247*, 91–101.

Kanehisa, M., Sato, Y., and Morishima, K. (2016). BlastKOALA and GhostKOALA: KEGG tools for functional characterization of genome and metagenome sequences. J. Mol. Biol. *428*, 726–731.

Karsenti, E., Acinas, S.G., Bork, P., Bowler, C., De Vargas, C., Raes, J., Sullivan, M., Arendt, D., Benzoni, F., Claverie, J.M., *et al.* (2011). A holistic approach to marine eco-systems biology. PLOS Biol. *9*, e1001177. https://doi.org/10.1371/journal.pbio.1001177

Katoh, K., Rozewicki, J., and Yamada, K.D. (2017). MAFFT online service: multiple sequence alignment, interactive sequence choice and visualization. Brief. Bioinformatics [Epub ahead of print]. https://doi.org/10.1093/bib/bbx108

Kelley, L.A., Mezulis, S., Yates, C.M., Wass, M.N., and Sternberg, M.J. (2015). The Phyre2 web portal for protein modeling, prediction and analysis. Nat. Protoc. *10*, 845–858. https://doi.org/10.1038/nprot.2015.053

Kim, J.Y., Jo, B.H., Jo, Y., and Cha, H.J. (2012). Improved production of biohydrogen in light-powered *Escherichia coli* by co-expression of proteorhodopsin and heterologous hydrogenase. Microb. Cell Fact. *11*, 2. https://doi.org/10.1186/1475-2859-11-2

Kim, K., Kwon, S.K., Jun, S.H., Cha, J.S., Kim, H., Lee, W., Kim, J.F., and Cho, H.S. (2016). Crystal structure and functional characterization of a light-driven chloride pump having an NTQ motif. Nat. Commun. *7*, 12677. https://doi.org/10.1038/ncomms12677

Klöser, H., Ferreyra, G., Schloss, I., Mercuri, G., Laturnus, F., and Curtosi, A. (1994). Hydrography of Potter Cove, a small fjord-like inlet on King George island (South Shetlands). Estuar. Coast. Shelf Sci. *38*, 523–537.

Klotz, M.G., and Stein, L.Y. (2008). Nitrifier genomics and evolution of the nitrogen cycle. FEMS Microbiol. Lett. *278*, 146–156.

Klotz, M.G., Schmid, M.C., Strous, M., op den Camp, H.J., Jetten, M.S., and Hooper, A.B. (2008). Evolution of an octahaem cytochrome c protein family that is key to aerobic and anaerobic ammonia oxidation by bacteria. Environ. Microbiol. *10*, 3150–3163. https://doi.org/10.1111/j.1462-2920.2008.01733.x

Könneke, M., Bernhard, A.E., de la Torre, J.R., Walker, C.B., Waterbury, J.B., and Stahl, D.A. (2005). Isolation of an autotrophic ammonia-oxidizing marine archaeon. Nature *437*, 543–546.

Kopf, A., Bicak, M., Kottmann, R., Schnetzer, J., Kostadinov, I., Lehmann, K., Fernandez-Guerra, A., Jeanthon, C., Rahav, E., Ullrich, M., *et al.* (2015). The ocean sampling day consortium. Gigascience *4*, 27. https://doi.org/10.1186/s13742-015-0066-5

Kozlikova, B., Sebestova, E., Sustr, V., Brezovsky, J., Strnad, O., Daniel, L., Bednar, D., Pavelka, A., Manak, M., Bezdeka, M., *et al.* (2014). CAVER Analyst 1.0: graphic tool for interactive visualization and analysis of tunnels and channels in protein structures. Bioinformatics *30*, 2684–2685. https://doi.org/10.1093/bioinformatics/btu364

Krieger, E., Nabuurs, S.B., and Vriend, G. (2005). Homology Modeling. In Structural Bioinformatics, Gu, J., Bourne, P.E., eds. (John Wiley & Sons, Inc., Hoboken, NJ), pp. 509–523.

Krogh, A., Larsson, B., von Heijne, G., and Sonnhammer, E.L. (2001). Predicting transmembrane protein topology with a hidden Markov model: application to complete genomes. J. Mol. Biol. *305*, 567–580. https://doi.org/10.1006/jmbi.2000.4315

Lacroix, T., Loux, V., Gendrault, A., Hoebeke, M., and Gibrat, J.-F. (2014). Insyght: navigating amongst abundant homologues, syntenies and gene functional annotations in bacteria, it's that symbol! Nucleic Acids Res. *42*, e162.

Laskowski, R.A., MacArthur, M.W., Moss, D.S., and Thornton, J.M. (1993). PROCHECK: a program to check the stereochemical quality of protein structures. J. Appl. Crystallogr. *26*, 283–291.

Lee, K.A., Lee, S.S., Kim, S.Y., Choi, A.R., Lee, J.H., and Jung, K.H. (2015). Mistic-fused expression of algal rhodopsins in *Escherichia coli* and its photochemical properties. Biochim. Biophys. Acta *1850*, 1694–1703. https://doi.org/10.1016/j.bbagen.2015.04.002

Legendre, P., and Legendre, L.F.J. (1998). Numerical Ecology (Elsevier, Oxford).

Lidstrom, M.E. (2006). Aerobic methylotrophic prokaryotes. In The Prokaryotes, Dworkin, M., Falkrow, S., Rosemberg, E., Schleifer, K-H., and Stackebrandt, E., eds. (Springer, New York), pp. 618–634.

Lieberman, D., Lieberman, M., Peralta, R., and Hartshorn, G.S. (1996). Tropical forest structure and composition on a large-scale altitudinal gradient in Costa Rica. J. Ecol. *84*, 137–152.

López, J.L., Golemba, M., Hernández, E., Lozada, M., Dionisi, H., Jansson, J.K., Carroll, J., Lundgren, L., Sjöling, S., and Mac Cormack, W.P. (2017). Microbial and viral-like rhodopsins present in coastal marine sediments from four polar and subpolar regions. FEMS Microbiol. Ecol. *93*,. https://doi.org/10.1093/femsec/fiw216

Loviso, C.L., Lozada, M., Guibert, L.M., Musumeci, M.A., Sarango Cardenas, S., Kuin, R.V., Marcos, M.S., and Dionisi, H.M. (2015). Metagenomics reveals the high polycyclic aromatic hydrocarbon-degradation potential of abundant uncultured bacteria from chronically polluted subantarctic and temperate coastal marine environments. J. Appl. Microbiol. *119*, 411–424. https://doi.org/10.1111/jam.12843

Liuzzi, M.G., Gappa, J.L., and Piriz, M.L. (2011). Latitudinal gradients in macroalgal biodiversity in the Southwest Atlantic between 36 and 55°S. Hydrobiologia *673*, 205–214.

Lozupone, C.A., and Knight, R. (2007). Global patterns in bacterial diversity. Proc. Natl. Acad. Sci. U.S.A. *104*, 11436–11440.

Marcos, M.S., Lozada, M., Di Marzio, W.D., and Dionisi, H.M. (2012). Abundance, dynamics, and biogeographic distribution of seven polycyclic aromatic hydrocarbon dioxygenase gene variants in coastal sediments of Patagonia. Appl. Environ. Microbiol. *78*, 1589–1592. https://doi.org/10.1128/AEM.06929-11

Mascotti, M.L., Lapadula, W.J., and Juri Ayub, M. (2015). The origin and evolution of Baeyer-Villiger monooxygenases (BVMOs): an ancestral family of flavin monooxygenases. PLOS ONE *10*, e0132689. https://doi.org/10.1371/journal.pone.0132689

Mason, O.U., Scott, N.M., Gonzalez, A., Robbins-Pianka, A., Bælum, J., Kimbrel, J., Bouskill, N.J., Prestat, E., Borglin, S., Joyner, D.C., *et al.* (2014). Metagenomics reveals sediment microbial community response to Deepwater Horizon oil spill. ISME J. *8*, 1464–1475. https://doi.org/10.1038/ismej.2013.254

Matos, M.N., Lozada, M., Anselmino, L.E., Musumeci, M.A., Henrissat, B., Jansson, J.K., Mac Cormack, W.P., Carroll, J., Sjöling, S., Lundgren, L., *et al.* (2016). Metagenomics unveils the attributes of the alginolytic guilds of sediments from four distant cold coastal environments. Environ. Microbiol. *18*, 4471–4484. https://doi.org/10.1111/1462-2920.13433

Matsen, F.A., Kodner, R.B., and Armbrust, E.V. (2010). pplacer: linear time maximum-likelihood and Bayesian phylogenetic placement of sequences onto a fixed reference tree. BMC Bioinformatics *11*, 538. https://doi.org/10.1186/1471-2105-11-538

Minic, J., Grosclaude, J., Aioun, J., Persuy, M.A., Gorojankina, T., Salesse, R., Pajot-Augy, E., Hou, Y., Helali, S., Jaffrezic-Renault, N., *et al.* (2005). Immobilization of native membrane-bound rhodopsin on biosensor surfaces. Biochim. Biophys. Acta *1724*, 324–332.

Mojica, K.D., and Brussaard, C.P. (2014). Factors affecting virus dynamics and microbial host-virus interactions in marine environments. FEMS Microbiol. Ecol. *89*, 495–515. https://doi.org/10.1111/1574-6941.12343

Mountfort, D.O., Rainey, F.A., Burghardt, J., Kaspar, H.F., and Stackebrandt, E. (1998). *Psychromonas antarcticus* gen. nov., sp. nov., a new aerotolerant anaerobic, halophilic psychrophile isolated from pond sediment of the McMurdo Ice Shelf, Antarctica. Arch. Microbiol. *169*, 231–238.

Muegge, B.D., Kuczynski, J., Knights, D., Clemente, J.C., González, A., Fontana, L., Henrissat, B., Knight, R., and Gordon, J.I. (2011). Diet drives convergence in gut microbiome functions across mammalian phylogeny and within humans. Science *332*, 970–974. https://doi.org/10.1126/science.1198719

Mukherjee, S., Seshadri, R., Varghese, N.J., Eloe-Fadrosh, E.A., Meier-Kolthoff, J.P., Göker, M., Coates, R.C., Hadjithomas, M., Pavlopoulos, G.A., Paez-Espino, D., *et al.* (2017). 1,003 reference genomes of bacterial and archaeal isolates expand coverage of the tree of life. Nat. Biotechnol. *35*, 676–683. https://doi.org/10.1038/nbt.3886

Musumeci, M.A., Lozada, M., Rial, D.V., Mac Cormack, W.P., Jansson, J.K., Sjöling, S., Carroll, J., and Dionisi, H.M. (2017). Prospecting biotechnologically-relevant monooxygenases from cold sediment metagenomes: an in silico approach. Mar. Drugs *15*, E114.

Naidu, M.T., and Kumar, O.A. (2016). Tree diversity, stand structure, and community composition of tropical forests in Eastern Ghats of Andhra Pradesh, India. J. Asia-Pac. Biodivers. *9*, 328–334.

Nealson, K.H. (1997). Sediment bacteria: who's there, what are they doing, and what's new? Annu. Rev. Earth Planet. Sci. *25*, 403–434. https://doi.org/10.1146/annurev.earth.25.1.403

Neumann, A.M., Balmonte, J.P., Berger, M., Giebel, H.A., Arnosti, C., Voget, S., Simon, M., Brinkhoff, T., and Wietz, M. (2015). Different utilization of alginate and other algal polysaccharides by marine *Alteromonas macleodii* ecotypes. Environ. Microbiol. *17*, 3857–3868. https://doi.org/10.1111/1462-2920.12862

Nie, Y., Chi, C.Q., Fang, H., Liang, J.L., Lu, S.L., Lai, G.L., Tang, Y.Q., and Wu, X.L. (2014). Diverse alkane hydroxylase genes in microorganisms and environments. Sci. Rep. *4*, 4968. https://doi.org/10.1038/srep04968

Nogi, Y., Kato, C., and Horikoshi, K. (2002). *Psychromonas kaikoae* sp. nov., a novel from the deepest piezophilic bacterium cold-seep sediments in the Japan Trench. Int. J. Syst. Evol. Microbiol. *52*, 1527–1532.

Norton, J.M., Alzerreca, J.J., Suwa, Y., and Klotz, M.G. (2002). Diversity of ammonia monooxygenase operon in autotrophic ammonia-oxidizing bacteria. Arch. Microbiol. *177*, 139–149. https://doi.org/10.1007/s00203-001-0369-z

Parker, J., Rambaut, A., and Pybus, O.G. (2008). Correlating viral phenotypes with phylogeny: accounting for phylogenetic uncertainty. Infect. Genet. Evol. *8*, 239–246.

Pettersen, E.F., Goddard, T.D., Huang, C.C., Couch, G.S., Greenblatt, D.M., Meng, E.C., and Ferrin, T.E. (2004). UCSF Chimera – a visualization system for exploratory research and analysis. J. Comput. Chem. *25*, 1605–1612. https://doi.org/10.1002/jcc.20084

Philosof, A., and Béjà, O. (2013). Bacterial, archaeal and viral-like rhodopsins from the Red Sea. Environ. Microbiol. Rep. *5*, 475–482. https://doi.org/10.1111/1758-2229.12037

Quartino, M.L., and Boraso de Zaixso, A.L. (2008). Summer macroalgal biomass in Potter Cove, South Shetland Islands, Antarctica: its production and flux to the ecosystem. Polar Biol. *31*, 281–294.

Rabus, R., Boll, M., Heider, J., Meckenstock, R.U., Buckel, W., Einsle, O., Ermler, U., Golding, B.T., Gunsalus, R.P., Kroneck, P.M., *et al.* (2016). Anaerobic microbial degradation of hydrocarbons: from enzymatic reactions to the environment. J. Mol. Microbiol. Biotechnol. *26*, 5–28. https://doi.org/10.1159/000443997

Ramette, A. (2007). Multivariate analyses in microbial ecology. FEMS Microbiol. Ecol. *62*, 142–160.

Ravishankara, A.R., Daniel, J.S., and Portmann, R.W. (2009). Nitrous oxide (N2O): the dominant ozone-depleting substance emitted in the 21st century. Science *326*, 123–125. https://doi.org/10.1126/science.1176985

Reyes, C., Schneider, D., Lipka, M., Thürmer, A., Böttcher, M.E., and Friedrich, M.W. (2017). Nitrogen metabolism genes from temperate marine sediments. Mar. Biotechnol. *19*, 175–190. https://doi.org/10.1007/s10126-017-9741-0

Rideout, J.R., He, Y., Navas-Molina, J.A., Walters, W.A., Ursell, L.K., Gibbons, S.M., Chase, J., McDonald, D., Gonzalez, A., Robbins-Pianka, A., *et al.* (2014). Subsampled open-reference clustering creates consistent, comprehensive OTU definitions and scales to billions of sequences. PeerJ *2*, e545. https://doi.org/10.7717/peerj.545

Riebel, A., Dudek, H.M., de Gonzalo, G., Stepniak, P., Rychlewski, L., and Fraaije, M.W. (2012). Expanding the set of rhodococcal Baeyer-Villiger monooxygenases by high-throughput cloning, expression and substrate screening. Appl. Microbiol. Biotechnol. *95*, 1479–1489. https://doi.org/10.1007/s00253-011-3823-0

Riesenfeld, C.S., Schloss, P.D., and Handelsman, J. (2004). Metagenomics: genomic analysis of microbial communities. Annu. Rev. Genet. *38*, 525–552. https://doi.org/10.1146/annurev.genet.38.072902.091216

Rocha, E.P. (2006). Inference and analysis of the relative stability of bacterial chromosomes. Mol. Biol. Evol. *23*, 513–522.

Ronquist, F., and Huelsenbeck, J.P. (2003). MrBayes 3: Bayesian phylogenetic inference under mixed models. Bioinformatics *19*, 1572–1574.

Rotthauwe, J.H., Witzel, K.P., and Liesack, W. (1997). The ammonia monooxygenase structural gene amoA as a functional marker: molecular fine-scale analysis of natural ammonia-oxidizing populations. Appl. Environ. Microbiol. *63*, 4704–4712.

Sabehi, G., Massana, R., Bielawski, J.P., Rosenberg, M., Delong, E.F., and Béjà, O. (2003). Novel Proteorhodopsin variants from the Mediterranean and Red Seas. Environ. Microbiol. *5*, 842–849.

Salazar, G., and Sunagawa, S. (2017). Marine microbial diversity. Curr. Biol. *27*, R489–R494.

Schloss, P.D., and Westcott, S.L. (2011). Assessing and improving methods used in operational taxonomic unit-based approaches for 16S rRNA gene sequence analysis. Appl. Environ. Microbiol. *77*, 3219–3226. https://doi.org/10.1128/AEM.02810-10

Schmid, M.C., Hooper, A.B., Klotz, M.G., Woebken, D., Lam, P., Kuypers, M.M., Pommerening-Roeser, A., Op den Camp, H.J., and Jetten, M.S. (2008). Environmental detection of octahaem cytochrome c hydroxylamine/hydrazine oxidoreductase genes of aerobic and anaerobic ammonium-oxidizing bacteria. Environ. Microbiol. *10*, 3140–3149. https://doi.org/10.1111/j.1462-2920.2008.01732.x

Sessitsch, A., Hardoim, P., Döring, J., Weilharter, A., Krause, A., Woyke, T., Mitter, B., Hauberg-Lotte, L., Friedrich, F., Rahalkar, M., *et al.* (2012). Functional characteristics of an endophyte community colonizing rice roots as revealed by metagenomic analysis. Mol. Plant Microbe Interact. *25*, 28–36. https://doi.org/10.1094/MPMI-08-11-0204

Slamovits, C.H., Okamoto, N., Burri, L., James, E.R., and Keeling, P.J. (2011). A bacterial proteorhodopsin proton pump in marine eukaryotes. Nat. Commun. *2*, 183. https://doi.org/10.1038/ncomms1188

Snelgrove, P.V. (1999). Getting to the bottom of marine biodiversity: sedimentary habitats: ocean bottoms are the most widespread habitat on earth and support high biodiversity and key ecosystem services. BioScience *49*, 129–138.

Spalding, M.D., Fox, H.E., Allen, G.R., Davidson, N., Ferdaña, Z.A., Finlayson, M., Halpern, B.S., Jorge, M.A., Lombana, A., Lourie, S.A., *et al.* (2007). Marine ecoregions of the world: a bioregionalization of coastal and shelf areas. BioScience *57*, 573–583.

Spudich, J.L. (2006). The multitalented microbial sensory rhodopsins. Trends Microbiol. *14*, 480–487.

Spudich, J.L., Yang, C.S., Jung, K.H., and Spudich, E.N. (2000). Retinylidene proteins: structures and functions from archaea to humans. Annu. Rev. Cell Dev. Biol. *16*, 365–392. https://doi.org/10.1146/annurev.cellbio.16.1.365

Stackebrandt, E., Frederiksen, W., Garrity, G.M., Grimont, P.A., Kampfer, P., Maiden, M.C., Nesme, X., Rossello-Mora, R., Swings, J., Truper, H.G., *et al.* (2002). Report of the *ad hoc* committee for the re-evaluation of the species definition in bacteriology. Int. J. Syst. Evol. Microbiol. *52*, 1043–1047.

Stamatakis, A. (2014). RAxML version 8: a tool for phylogenetic analysis and post-analysis of large phylogenies. Bioinformatics *30*, 1312–1313. https://doi.org/10.1093/bioinformatics/btu033

Stein, L.Y. (2011). Heterotrophic nitrification and nitrifier denitrification. In Nitrification, Ward, B., Arp, D., and Klotz, M.G., eds. (ASM Press, Washington, DC), pp. 95–114.

Stein, L.Y., and Klotz, M.G. (2011). Nitrifying and denitrifying pathways of methanotrophic bacteria. Biochem. Soc. Trans. *39*, 1826–1831. https://doi.org/10.1042/BST20110712

Steinle, L., Maltby, J., Treude, T., Kock, A., Bange, H.W., Engbersen, N., Zopfi, J., Lehmann, M.F., and Niemann, H. (2017). Effects of low oxygen concentrations on aerobic methane oxidation in seasonally hypoxic coastal waters. Biogeosciences BG *14*, 1631–1645.

Strous, M., Fuerst, J.A., Kramer, E.H., Logemann, S., Muyzer, G., van de Pas-Schoonen, K.T., Webb, R., Kuenen, J.G., and Jetten, M.S. (1999). Missing lithotroph identified as new planctomycete. Nature *400*, 446–449. https://doi.org/10.1038/22749

Strous, M., Pelletier, E., Mangenot, S., Rattei, T., Lehner, A., Taylor, M.W., Horn, M., Daims, H., Bartol-Mavel, D., Wincker, P., *et al.* (2006). Deciphering the evolution and metabolism of an anammox bacterium from a community genome. Nature *440*, 790–794.

Sul, W.J., Oliver, T.A., Ducklow, H.W., Amaral-Zettler, L.A., and Sogin, M.L. (2013). Marine bacteria exhibit a bipolar distribution. Proc. Natl. Acad. Sci. U.S.A. *110*, 2342–2347. https://doi.org/10.1073/pnas.1212424110

Takagi, T., Morisaka, H., Aburaya, S., Tatsukami, Y., Kuroda, K., and Ueda, M. (2016). putative alginate assimilation process of the marine bacterium *Saccharophagus degradans* 2-40 based on quantitative proteomic analysis. Mar. Biotechnol. *18*, 15–23. https://doi.org/10.1007/s10126-015-9667-3

Tatusov, R.L., Galperin, M.Y., Natale, D.A., and Koonin, E.V. (2000). The COG database: a tool for genome-scale analysis of protein functions and evolution. Nucleic Acids Res. *28*, 33–36.

Terrapon, N., Lombard, V., Drula, E., Coutinho, P.M., and Henrissat, B. (2017). The CAZy database/the carbohydrate-active enzyme (CAZy) database: principles and usage guidelines. In A Practical Guide to Using Glycomics Databases, Aoki-Kinoshita, K.F., ed. (Springer Japan, Tokyo), pp. 117–131.

Thomas, F., Barbeyron, T., Tonon, T., Génicot, S., Czjzek, M., and Michel, G. (2012). Characterization of the first alginolytic operons in a marine bacterium: from their emergence in marine Flavobacteriia to their independent transfers to marine Proteobacteria and human gut Bacteroides. Environ. Microbiol. *14*, 2379–2394. https://doi.org/10.1111/j.1462-2920.2012.02751.x

Thomas, F., Bordron, P., Eveillard, D., and Michel, G. (2017). Gene expression analysis of *Zobellia galactanivorans* during the degradation of algal polysaccharides reveals both substrate-specific and shared transcriptome-wide responses. Front. Microbiol. *8*, 1808. https://doi.org/10.3389/fmicb.2017.01808

Thompson, J.D., Gibson, T.J., and Higgins, D.G. (2002). Multiple sequence alignment using ClustalW and ClustalX. Curr. Protoc. Bioinformatics *Chapter*, Unit 2.3. https://doi.org/10.1002/0471250953.bi0203s00

Thompson, L.R., Sanders, J.G., McDonald, D., Amir, A., Ladau, J., Locey, K.J., Prill, R.J., Tripathi, A., Gibbons, S.M., Ackermann, G., *et al.* (2017). A communal catalogue reveals Earth's multiscale microbial diversity. Nature *551*, 457–463. https://doi.org/10.1038/nature24621

Touchon, M., and Rocha, E.P. (2016). Coevolution of the Organization and Structure of Prokaryotic Genomes. Cold Spring Harb. Perspect. Biol. *8*, a018168. https://doi.org/10.1101/cshperspect.a018168

Trott, O., and Olson, A.J. (2010). AutoDock Vina: improving the speed and accuracy of docking with a new scoring function, efficient optimization, and multithreading. J. Comput. Chem. *31*, 455–461. https://doi.org/10.1002/jcc.21334

Tsunoda, S.P., Prigge, M., Abe-Yoshizumi, R., Inoue, K., Kozaki, Y., Ishizuka, T., Yawo, H., Yizhar, O., and Kandori, H. (2017). Functional characterization of sodium-pumping rhodopsins with different pumping properties. PLOS ONE *12*, e0179232. https://doi.org/10.1371/journal.pone.0179232

Turner, J., Bindschadler, R., Convey, P., DiPrisco, G., Fahrbach, E., Gutt, J., Hodgson, D., Mayewski, P., and Summerhayes, C. (2009). Antarctic climate change and the environment: a contribution to the International Polar Year 2007–2008 (Scientific Committee on Antarctic Research, Cambridge).

van Beilen, J.B., and Funhoff, E.G. (2007). Alkane hydroxylases involved in microbial alkane degradation. Appl. Microbiol. Biotechnol. *74*, 13–21. https://doi.org/10.1007/s00253-006-0748-0

Wang, Y., Sheng, H.F., He, Y., Wu, J.Y., Jiang, Y.X., Tam, N.F., and Zhou, H.W. (2012). Comparison of the levels of bacterial diversity in freshwater, intertidal wetland, and marine sediments by using millions of illumina tags. Appl. Environ. Microbiol. *78*, 8264–8271. https://doi.org/10.1128/AEM.01821-12

Webb, B., and Sali, A. (2014). Protein structure modeling with MODELLER. In Protein Structure Prediction, Kihara, D., eds. (Humana Press, New York), pp. 1–15

Wöhlbrand, L., Jacob, J.H., Kube, M., Mussmann, M., Jarling, R., Beck, A., Amann, R., Wilkes, H., Reinhardt, R., and Rabus, R. (2013). Complete genome, catabolic sub-proteomes and key-metabolites of *Desulfobacula toluolica* Tol2, a marine, aromatic compound-degrading, sulfate-reducing bacterium. Environ. Microbiol. *15*, 1334–1355. https://doi.org/10.1111/j.1462-2920.2012.02885.x

Yoon, J. (2014). The family Verrucomicrobiaceae. In The Prokaryotes, Rosenberg, E., DeLong, E.F., Lory, S., Stackebrandt, E., and Thompson, F., eds. (Springer, Berlin Heidelberg), pp. 1017–1018.

Yu, K., and Zhang, T. (2012). Metagenomic and metatranscriptomic analysis of microbial community structure and gene expression of activated sludge. PLOS ONE *7*, e38183. https://doi.org/10.1371/journal.pone.0038183

Yutin, N., and Koonin, E.V. (2012). Proteorhodopsin genes in giant viruses. Biol. Direct *7*, 34. https://doi.org/10.1186/1745-6150-7-34

Zehr, J.P., and Kudela, R.M. (2011). Nitrogen cycle of the open ocean: from genes to ecosystems. Ann. Rev. Mar. Sci. *3*, 197–225. https://doi.org/10.1146/annurev-marine-120709-142819

Zhu, B., and Yin, H. (2015). Alginate lyase: Review of major sources and classification, properties, structure-function analysis and applications. Bioengineered *6*, 125–131. https://doi.org/10.1080/21655979.2015.1030543

Wildlife Microbial Genomics and Endocrinology

6

Holly L. Lutz[1,2*], Sophia Carryl[3,4] and
Rachel M. Santymire[3,4]

[1]The Microbiome Center, Department of Surgery, University of Chicago, Chicago, IL, USA.
[2]Integrative Research Center, Field Museum of Natural History, Chicago, IL, USA.
[3]Committee on Evolutionary Biology, University of Chicago, Chicago, IL, USA.
[4]Davee Center for Epidemiology and Endocrinology, Lincoln Park Zoo, Chicago, IL, USA.

*Correspondence: hlutz@fieldmuseum.org

https://doi.org/10.21775/9781912530021.06

Abstract

Microbes play an integral role in the healthy functioning of most living systems and a disturbed microbiome can be an indication of – or invitation for – infection and disease. In an era of accelerated anthropogenic change and environmental disturbance, the biological characterization of natural systems is critical. The advent of 'omics' approaches to microbial ecology has provided a new means for these systems to be described and is leading to unprecedented insights into organismal health and ecosystem quality. The ability to characterize microbial variation within individuals and across species and populations has also provided a new framework in which to study symbiotic relationships. Indeed, the synthesis of microbial ecology and host and environmental biology provides the foundation for what is now considered the 'one health' approach to zoonotic disease surveillance and prevention. In addition to improving our basic knowledge of what constitutes a healthy organism, microbial 'omics' approaches can inform wildlife conservation and captive animal husbandry practices by providing preclinical indicators of stress, reproductive ability, and disease. When applied to natural history surveys of organisms and ecosystems, microbial genomics approaches can provide important baseline data for bacteria and viruses with zoonotic potential. The aim of this chapter is to provide an overview of how genomic approaches to microbial ecology can be paired with host biology, endocrinology, and disease data to characterize overall health of animals, habitats, and ecosystems and inform conservation and disease surveillance strategies.

Introduction

With the advent of new 'omics' approaches, it is now possible to not only consider eukaryotic organisms, but also to effectively assay the abundant symbiotic microbes (bacteria, fungi, archaea, and viruses) that they host. These symbionts, collectively referred to as the 'microbiome', represent a novel source of ecological and evolutionary data for the study of wildlife. The ratio of bacterial to host cells in humans ranges from 1.3 to 2.3 (Sender *et al.*, 2016), and it is now well understood that bacteria are essential to the body's proper functioning. Indeed, many diseases have now been linked to bacterial dysbiosis in the human gut, skin, and reproductive tract (Nicholson, 2012; Williams and Gallo, 2017; Chen *et al.*, 2017). In a sense, the microbiome functions as an additional organ in many eukaryotes – it can serve as a buffer to exogenous assaults, aid in digestion and nutrient acquisition, and prime the adaptive immune system.

With the establishment of adequate baseline data, symbionts may also serve as biomarkers for the overall health of the host organism. The use of genomic and metagenomic approaches via high-throughput sequencing platforms (e.g. Illumina, PacBio) is allowing researchers to assess both bacterial community composition and the relative abundance of operational taxonomic units (OTUs), as well as the putative function of these symbionts comprising a host's microbiome. These data, paired with physiological, ecological, and evolutionary data of host organisms, are providing important context for defining the 'health' of wildlife species. Furthermore, microbiome data can be paired with host genomic data (considered a 'hologenomic' approach) – and in some cases population-level single nucleotide polymorphism (SNP) databases – to identify links between host–symbiont associations and adaptive response (e.g. Rothschild, 2018).

Novel 'omics' approaches provide important tools for integrating the complex aspects of wildlife and ecosystem health, and these tools come at a critical time of unprecedented and broad-scale environment change. Biodiversity is currently threatened by a seemingly inexorable 'sixth mass extinction' event, referred to by many scientists as the 'Anthropocene Epoch', and characterized by significant human impact on the earth's geology and natural resources (Ceballos *et al.*, 2015). Anthropogenic disturbance includes, but is not limited to, deforestation, pollution, and exploitation of limited natural resources. Cultivated lands and domesticated animals are also maintained in precarious states of monoculture, modification, and over-prescription of antibiotics – the last of which is leading to bacterial 'extinctions' in many domesticated animals. Such disturbances can negatively impact both humans and wildlife (Mace *et al.*, 2012; Ehrlich and Ehrlich, 2013), and new approaches to quantification and amelioration of disturbance effects are in great demand. Efforts to conserve threatened wildlife species have included capturing and relocating animals to protected areas, and maintaining animals in captive enclosures such as zoos, aquaria, and reserves. National parks and wildlife reserves provide important areas of protected natural habitat for many animals and plants, while zoos serve an important role in conserving genetic diversity and promoting public education on wildlife (Moss *et al.*, 2015). Many measures are taken to ensure the health of individuals within these protected areas, but to date, conservation research has focused primarily on detecting diseases and documenting population growth and decline, having few means to identify mechanisms underlying successful reproduction or immunity to diseases.

Fortunately, conservation and captive animal husbandry research are beginning to experience a paradigm shift towards incorporating symbiotic organisms into the overall assessment of animal health (e.g. Bahrndorff *et al.*, 2016; DeCandia *et al.*, 2018). This holistic approach to animal health has a number of important benefits. First, it is driving the collection of data that are integral to wildlife biology, but that have often been overlooked.

Second, these data are leading to a finer resolution of habitat quality and ecosystem health. Third, these data can be applied to conservation efforts, particularly in the context of husbandry for wild species, for which proper habitat and diet can be difficult to provide. Finally, by taking a systematic and holistic approach to the study of microbial symbionts in wildlife, researchers are beginning to elucidate mechanisms by which pathogen transmission and disease are propagated. These benefits, taken together, constitute an important component of the 'One Health Initiative', defined by the US Centers for Disease Control and Prevention as '[recognizing] that the health of people is connected to the health of animals and the environment' (www.cdc.gov/onehealth/index.html) (Fig. 6.1).

Previously, most microbial research focused on elucidating the role of bacteria and viruses in pathogenesis and disease, without acknowledging any potential benefits provided by these symbionts to their hosts. Many studies are now demonstrating the important role that microbial communities serve in modulating, producing, and secreting hormones that directly affect the host's behaviour, immunity, reproduction, and metabolic function, as well as host growth and development (reviewed in Neuman *et al.*, 2015). Overlaying glucocorticoid (GC) analysis, microbial communities and host behaviour may be an improved perspective of health and fitness. In this chapter, we review the concept of animal 'health' and discuss mechanisms by which bacterial symbionts modulate animal behaviour. We then explore the nature of novel 'omics' approaches being applied to the study of microbes, and posit that the application of these tools to questions of wildlife microbial ecology is an issue of both conservation and ecosystem health.

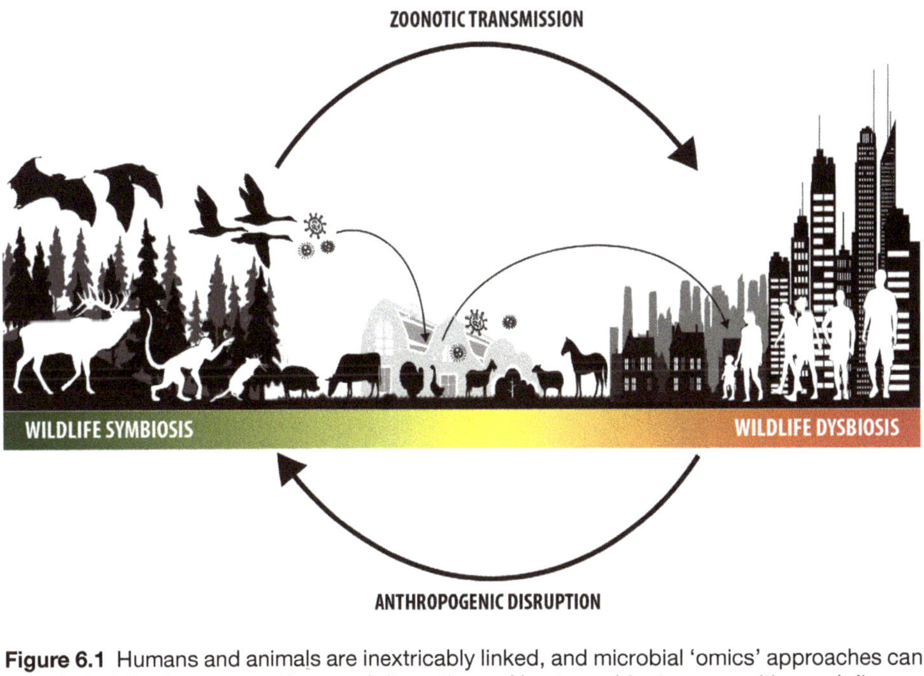

Figure 6.1 Humans and animals are inextricably linked, and microbial 'omics' approaches can provide insights into connections and disruptions of host–symbiont communities and disease transmission. Anthropogenic disruptions can have powerful effects on natural ecosystems, and subsequent dysbiosis can lead to increased stress and viral shedding. These factors, paired with increased contact between wildlife and livestock (or wildlife and humans), are some of the leading causes of pathogen transmission and zoonoses.

Defining animal health

The role of stress and the endocrine response

In animals, as with humans, the term 'health' generally refers to an individual's physiological condition and disease status. Signs of health that can be measured include heart rate, blood pressure, food and water consumption, lameness, lethargy, behaviour, and excrement dysfunction (Ryser-Degiorgis, 2013). Additionally, the presence of foreign material within the body, such as bacteria or eukaryotic parasites, can be assessed. Although bacteria and other microscopic symbionts were historically thought to be associated with disease, it is now recognized that freedom from such symbionts may not actually be a healthy – or even an achievable – state (Stephen, 2014). Being free of disease is not the same as the complete absences of a pathogen, and it is important to consider the extent to which disease is not disrupting homeostasis and impacting the health of the individual (Stephen, 2014). A more accurate measure of health is therefore an individual's ability to maintain homeostasis during times of pressure (Ryser-Degiorgis, 2013). In wildlife, stress is defined as a disturbance in this homeostasis, and may be caused by a perceived or physical threat (Touma and Palme, 2005). An animal might be considered unhealthy when it cannot mount a protective response to a stressor, such as a disease challenge, and then succumbs to the disease (Huber, 2011; Ryser-Degiorgis, 2013). The ability to cope with stressors, also called allostasis, is thus one of the key measures of animal health.

Distress in wildlife (vs. eustress, which is a stimulating or 'good' stress that does not threaten survival) can be induced by many pressures, both natural and unnatural in origin. Natural stressors may include social interactions, such as disputes over limited resources including food, shelter, or mates. Parasites and pathogens may also constitute natural stressors, although their prevalence and pathogenicity can be exacerbated by anthropogenic activities (e.g. hunting and poaching, habitat encroachment, introduction of non-native species), which produce unnatural stressors. These stressors elicit behavioural and physiological reactions, stimulating the hypothalamus–pituitary–adrenal (HPA) axis, which we discuss further below (McEwen and Wingfield, 2003).

The endocrinological pathways responsible for allostasis have been well studied. Generally, a physical or psychological (perceived) stress factor elicits a physiological response, activating the HPA axis. Having been activated, the hypothalamus releases corticotropin-releasing hormone (CRH) and arginine vasopressin (AVP), which then act on the anterior pituitary gland to produce adrenocorticotrophic hormone (ACTH). ACTH travels through the bloodstream and causes the adrenal glands to release glucocorticoids (GC; cortisol and corticosterone) from the cortex. The GCs cause the body to release glucose, sending energy to the organs vital to dealing with the stressor, such as the brain and heart. GCs then have a negative feedback mechanism to return the body to homeostasis (Romero, 2004; Reeder, 2005). However, if the stressors result in allostatic overload, leading to repeated release of GCs (McEwen and Wingfield, 2003), the individual may experience detrimental health effects and ultimately impairment of the body (Sapolsky, 2001; McEwen and Wingfield, 2003; Romero, 2004; Reeder and Kramer, 2005; Tarlow and Blumstein, 2007). Allostatic overload overrides the negative feedback loop, the liver becomes overextended because of its role in gluconeogenesis, and bodily functions not needed for survival (e.g. reproduction, growth, and development) are suspended or suppressed (Wingfield and Romero, 2001; Boonstra, 2005; Wingfield, 2005).

Glucocorticoids exhibit steroidal anti-inflammatory properties that can weaken the immune system by decreasing cytokine production and suppressing white blood cells, as well as several interleukins (ILs) including IL-1 and IL-2 (Welsh *et al.*, 1999) that act on immune cell receptors (Besedovsky,1996). These anti-inflammatory properties may also lead to changes in immune gene expression in target tissues (Padgett, 2003), increasing susceptibility to disease. This allostatic state can weaken the body's defence mechanisms, increasing the individual's susceptibility to parasites and pathogens. Chronic or repeated stressors may also reduce cognitive abilities via neuron loss, result in muscle atrophy as a result of protein loss, alter behaviours affecting intra-group relationships, increase disease transmission, and make individuals targets of predation, all of which ultimately impact their fitness (Sapolsky, 2001; McEwen and Wingfield, 2003; Charmandari *et al.*, 2005; Hing, 2016). Increasing numbers of emerging infectious diseases in the last few decades indicate that anthropogenic pressures are affecting wildlife health (Harper and Austad, 2004), and natural resource managers are being increasingly tasked with providing causal mechanisms between land use changes and wildlife health (Baker *et al.*, 2013). Faced with these demands, new tools and refined measures of wildlife and ecosystem health are of great importance.

Measuring wildlife stress

Traditionally, wildlife GCs were measured in blood (Sapolsky, 2001). However, because of the propensity for GC to rapidly rise in the bloodstream of animals when captured, such measurements may be an invalid reflection of their response to the environment (Harper and Austad, 2000; Möstl and Palme, 2002; Millspaugh and Washburn, 2004; Touma and Palme, 2005). Rather, the hormone concentrations are more likely to be a reflection of the capture process, providing just a snapshot of hormonal activity (Monfort, 2003). Since the late 1980s, wildlife endocrinologist have relied on faeces to obtain a more holistic view of wildlife stress physiology. In addition to being a non-invasive sampling technique, there are several important advantages of using faeces for evaluating stress physiology. First, faecal material contains a pooled value of stress hormone concentrations, typically representing an average of hormonal activity from the last 12–24 hours. Second, faecal samples provide sufficient amounts of material for most hormonal enzyme quantification or radio immunoassay protocols, which generally require from 0.02 grams (Poessel *et al.*, 2011) to 0.5 grams (Santymire and Armstrong, 2010). A third advantage is that the animal does not need to be directly observed; the faecal sample simply needs to be located and obtained. For example, remote cameras placed on latrines have been used to obtain samples from known black rhinoceros individuals (*Diceros bicornis bicornis*; Freeman *et al.*, 2014a,b). Lastly, the production of faeces is a frequent and fairly consistent event, making faeces a reliable source of material for study. This regularity facilitates repeated sampling of individuals for longitudinal studies, which can provide insights into the nature of stress responses (e.g. acute vs. chronic responses).

As with humans, individual animals have distinct baseline levels of stress. Once collected, it is therefore important to verify that the quantity of GC or GC metabolites accurately reflects the animal's stress physiology. Factors that can affect GC production include taxonomy [e.g. the major GC is cortisol in most primates, carnivores, and ungulates, whereas corticosterone is the major GC in rodents, birds, and reptiles (Wasser *et al.*, 2000; Touma and Palme, 2005)], sex, age, metabolic rate, social pressures, nutritional factors, daily rhythms, seasonality, and individual variation (reviewed in Goymann, 2012). To account for natural variation in GC concentration, other measures of health and well-being need to be quantified simultaneously. Additional measures may include behavioural observations (passiveness, pacing, self-mutilation), physical measurements such as body condition, or parasite load (reviewed in Baker *et al.*, 2013).

Microbes, host ecology, and regulation of stress responses

Underlying mechanisms regulating behaviour were once thought to solely include neural anatomy, gene expression, and endocrinology (Martin, 2010). However, recent studies are uncovering the role of the host-associated microbiota in key biological processes that shape stress responses, revealing a hidden dimension in animal behaviour. This newly identified area of research, known as 'microbial endocrinology', is based on the recognition of bidirectional relationships between microbes and their hosts, where the microbiome influence the host and the host influence the microbiome (e.g. microbes control host stress response, and that stress response alters the host microbiome) (Ezenwa, 2012; Lyte, 2013). Microbes can affect the expression of host stress responses directly by producing hormones. For example, bacteria are capable of producing gamma-aminobutyric acid (GABA), which plays a key role in the HPA axis (Foster and McVey Neufeld, 2013). Microbes may also influence the host stress response and subsequent behaviours indirectly via modulation of the adrenal cortex function (Neuman *et al.*, 2015) (Fig. 6.2). It is important to note that microbial modulation

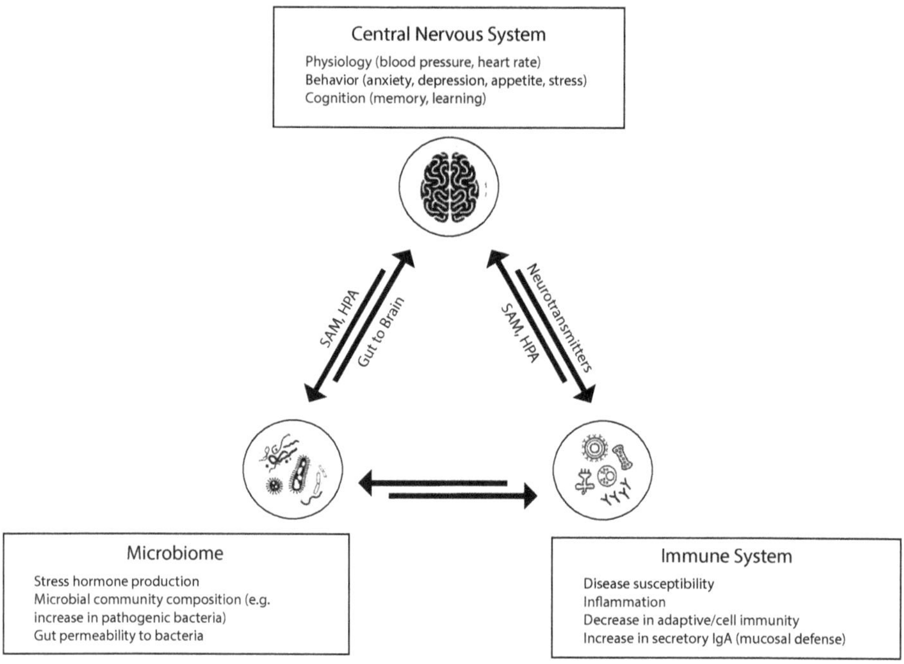

Figure 6.2 The stress response and its effects on physiology and behaviour are mediated by the interactions among the central nervous system, gut microbiota, and immune system. Perceived stress can trigger the sympathetic–adrenal medulla (SAM) pathway for fight or flight or the hypothalamus–pituitary-adrenal axis (HPA), which can trigger a response from both the microbiota and the immune system.

of stress responses via hormone production are non-infectious and independent of host immune functions (Sudo *et al.*, 2004; Diaz Heijtz, 2011; Lyte, 2013). The frequency and duration of stress responses can have implications for an animal's health, and reproductive success, making it of evolutionary importance to understand whether microbes can alter stress responses in their avian hosts.

Many mammalian studies have shown that the host microbiome can modulate the behavioural stress response. The bacterial genus *Lactobacillus* (phylum Firmicutes) includes a number of species capable of producing an inhibitory neurotransmitter, GABA. These species have been experimentally shown to decrease anxiety- and stress-like behaviours [reviewed in Neuman *et al.* (2015)]. Similarly, the bacterium *Bifidobacterium longum* was demonstrated to reduce levels of stress hormones and anxiety-like behaviours in both rats and humans (Neuman *et al.*, 2015). Moreover, GABA can dampen stress responses via its inhibitory effect on ACTH secretion, thus decreasing GC release (Míguez, 1990). Sudo *et al.* (2004) demonstrated increased stress reactivity in germ-free (GF) mice compared with specific pathogen-free (SPF) mice. Interestingly, when GF mice were conventionalized with the microbiome of SPF mice, anxiety-like behaviours were normalized. Additionally, oral administration of GABA-producing microbes (*Lactobacillus rhamnosus*) reduced anxiety- and stress-like behaviours in mice compared with controls. Mice administered with *L. rhamnosus* also showed reduced GC levels (Bravo, 2011; Boonstra *et al.*, 2015).

Fewer experimental studies have been conducted to examine the role of microbes in the stress response of birds, with most work focusing on pathogenicity of bacteria rather than host stress. Indeed, avian microbes have been best described in commercially bred poultry because of their economic importance (Stipkovits *et al.*, 1986; Lombardo, 1998). As a result, mostly pathogenic microbes were identified, including *Campylobacter* spp., *Clostridium botulinum* type C, *E. coli*, *Mycoplasma* spp., *Pasteurella multicolda*, *Pseudomonas* spp., *Salmonella* spp., *Shigella* spp., *Staphylococcus* spp., *Streptococcus* spp., *Vibrio* spp., and *Yersinia* spp. (Perek, 1969; Fritz, 1992; Lombardo, 1996, 1998). However, research has begun to expand beyond commercial animals and putatively pathogenic microbes. Many recent studies are now characterizing entire microbial communities in wild birds and other wildlife, and investigating the effect of these symbionts on host development, health, and physiology.

It is now understood that bacteria can serve as potential biomarkers for health and as important facilitators of behavioural responses in wildlife. For example, correlations between the prevalence of microbes (bacteria, fungi, and viruses) and mating behaviour have been documented in wild birds. Poiani and Wilks (2000) found that polygamous and polyandrous Australian passerines have greater microbial prevalence in their cloacae than monogamous species. They argued that acquiring pathogenic microbes via copulation modulates sexual promiscuity and therefore drives the evolution and maintenance of avian mating systems. Other studies support the aforementioned claim by providing evidence that exposure to new bacterial types is a consequence of having multiple mates, and that the potential gain of beneficial microbes can offset risks (Lombardo and Thorpe, 2000; Westneat, 2000; White *et al.*, 2010). Beneficial microbes can increase fitness in avian species in more direct ways as well. In African hoopoes (*Upupa epops*), breeding females secrete a microbial-rich fluid from a specialized gland known as the uropygial (preen) gland. The secretion has antibiotic-producing bacteria, *Enterococci*, that are active against Gram-positive and some Gram-negative bacteria, including *Bacillus licheniformis*, which is responsible for feather degradation. The oleaginous secretion is associated with hatching success in hoopoes and wood-hoopoes (*Phoeniculus purpureus*) as it decreases pathogenic microbial growth on the surface of eggshells, (Soler *et al.*, 2008; Martín-Vivaldi *et al.*, 2009; Ruiz-Rodríguez *et al.*, 2014). Similarly, house sparrows (*Passer domesticus*) with a larger uropygial gland show reduced feather wear, which supports prior studies that correlate large glands with greater microbial composition, thus inferring fitness benefits from microbial symbionts (Moreno-Rueda, 2010, 2011). In yet another example, researchers have discovered that pied flycatchers (*Ficedula hypoleuca*) host a bacterium, *Enterococcus fecium*, that is positively associated with fledgling mass and size, suggesting that *E. fecium* may act as a growth promoter by outcompeting pathogenic microbes (Moreno *et al.*, 2003).

Despite these examples of mutualistic relationships between hosts and their bacterial symbionts, the vast majority of relationships between wild animal hosts and their microbial symbionts remain uncharacterized and poorly understood. It is generally hypothesized that both host genetics and local environmental conditions shape bacterial community composition, and the role of symbionts as either helpful or harmful to host health is this likely to be context dependent, for example one bacterial species may provide benefits to the host under one set of conditions and may induce negative health effects under a separate set of conditions. To understand the conditions and contexts under which these relationships form and evolve, a large volume of data must be collected. Specifically, microbial communities living within wildlife hosts need to be characterized from a large number of individuals spanning geographical, ecological, and temporal ranges. Fortunately, the advent of high-throughput sequencing techniques has made this daunting task an achievable goal, as we discuss next.

'Omics' approaches to the study of wildlife microbial symbionts

Recent advances in molecular techniques – particularly nucleic acid-based approaches – have revolutionized the study of microbes. These advances have had profound impacts on our understanding of biodiversity, symbiotic relationships, and human and animal health. Traditional methods for the study of microbiology include microscopy and culturing techniques, and it is important to note that both still serve an important role in modern microbiological studies. However, with respect to identifying and detecting microbes from environmental samples, these traditional techniques are largely limited by the content of growth medium and the ability to provide and maintain appropriate conditions for bacterial, archaeal, or fungal growth (e.g. anaerobic chambers). The insensitivity of culturing methods has had major implications both for documenting microbial diversity and for identifying the presence or absence of clinically relevant microbes. Indeed, it is clear that insufficient culturing methods have left a large proportion of now-known microbial diversity in the dark (Thompson *et al.*, 2017). For many environments and organisms, it is estimated that ≈99% of microbial species cannot grow on laboratory media (Amann, 1995). Likewise, the inability of clinical researchers to reliably identify bacteria using culturing techniques has had significant impacts on patient health and even survival. The duration of time required by many specific bacterial cultures can be far greater than the amount of time a patient has to survive, as is the case with severe pneumonia infections (Kumar *et al.*, 2006; Laupland, 2013).

Technological advances and 'omics' approaches (i.e. culture-independent molecular techniques) are now driving more comprehensive analyses of microbial community composition. Specifically, advances in high-throughput sequencing and computing technologies have made it both logistically and financially feasible to accumulate large volumes of genetic sequence data from conserved 'barcoding' genes found in microbes. The most widely used bacterial and archaeal barcoding target is the ribosomal RNA gene (rRNA), but additional genomic loci exist. Shotgun metagenomic approaches are another option for studying archaea, fungi, and viruses, and provide important information about microbial community function in addition to taxonomic characterization (Kohl, 2012). We note that these advances in 'omics' approaches do not relegate traditional microbiological methods to obscurity, as these methods are still of great importance in understanding the basic function and biology of many microbes. Rather, these novel approaches complement traditional tools and provide a more effective means for comprehensively describing microbial diversity from a broad range of ecosystems.

Microbial symbionts persist in myriad anatomical and environmental sites, but regardless of their diverse origins, may be processed for sequencing using relatively similar methods. Biological samples collected for analysis of wildlife microbial symbionts typically include faecal or fluid material, localized swabs, or biopsied tissue. A growing body of evidence suggests that multiple methods of sample storage are sufficient for recovering microbial DNA; however, it is important to consider consistency of both sampling and downstream processing for comparative analyses (e.g. Song, 2016). Once collected, DNA can be extracted using a wide variety of methods. Many commercial kits exist specifically for microbial DNA extraction and, although yields *may* vary, most bead-beating and lysis-based kits are generally comparable (Ariefdjohan, 2010). Following DNA extraction, polymerase chain reaction (PCR) analysis is used to target and amplify the mitochondrial 16S rRNA 'barcoding' gene. This is followed by standard amplicon purification and quantitation methods, and subsequently sequencing on a high-throughput sequencing platform (e.g. Illumina HiSeq, San Diego, CA, USA). Finally, once sequence data have been obtained, there are a number of analytical pipelines and platforms with which bacterial (and archaeal) 16S rRNA data can be analysed to characterize the microbial community of each individual sample, host, etc. One currently popular pipeline is QIIME2™, an open-source next-generation bioinformatics platform that incorporates code curated by the platform's developers. In addition to 16S rRNA barcoding methods, entire shotgun metagenomic sequencing approaches may be utilized to not only describe the composition of a microbial community, but also examine the putative function of microbes in a given environment (e.g. Langille *et al.*, 2013). The field of microbial bioinformatics research is rapidly growing and thus it is, of course, prudent to familiarize oneself with the most recent advances in both DNA processing and bioinformatics analytical methods prior to conducting research in this area. We have provided here a general overview of the process for studying microbial diversity (outlined in Fig. 6.3).

The human–wildlife microbial interface: anthropogenic disruption of microbial ecology, consequences of dysbiosis, and the importance of a One Health approach

In light of the importance of bacteria and other microbial symbionts, awareness of adverse effects on wildlife microbiomes by anthropogenic modifications has increased. In invertebrate populations, such as honey bee colonies, exposure to commonly used pesticides has been shown to cause compositional and functional changes in their gut microbiota (Kakumanu *et al.*, 2016). Water contamination by heavy metals is now known to reduce both the

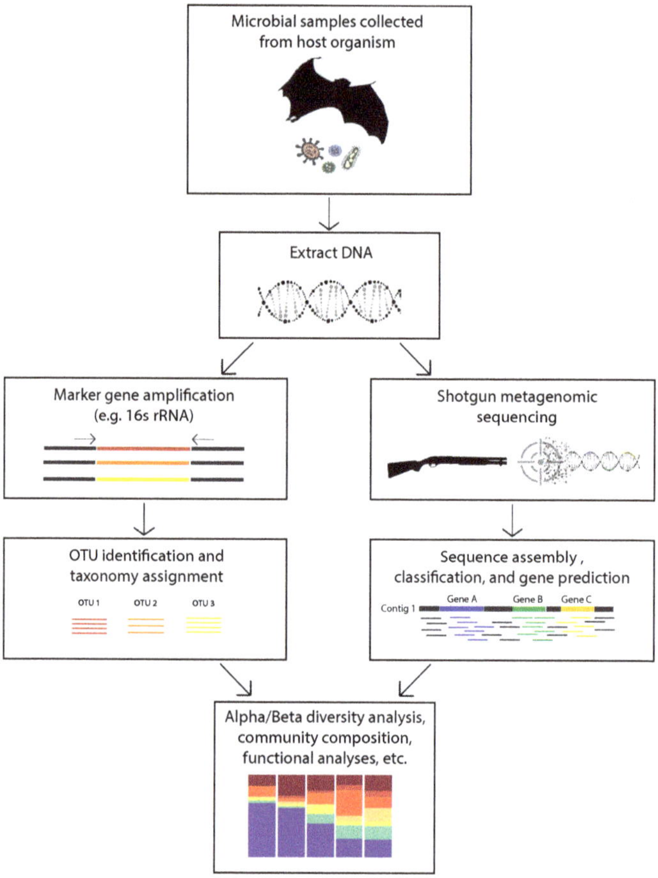

Figure 6.3 A general workflow for sampling and analysing microbial symbionts of animals (and other systems).

diversity and the abundance of the skin microbiota of frogs (Costa *et al.*, 2016), which may have serious implications for the spread of the catastrophic fungal disease, chytridiomycosis, caused by *Batrachochytrium dendobatidis* (Bd). Along similar lines, climate change was shown to considerably reduce gut microbiota diversity and alter composition and structure in lizards (Bestion, 2017), and habitat fragmentation proved to result in lower diversity and changes in taxonomic and functional composition in the gut microbiota of an African primate species (Barelli *et al.*, 2015). The impacts of anthropogenic disturbance are widespread and severe in many cases, and ignoring the significance of their effects on wildlife microbial ecology may lead to undesirable outcomes for both species conservation and human health.

In this chapter, we have discussed how endocrinological responses of animals to stress can, to some extent, be modulated by microbial symbionts. It follows that both host endocrinology and microbial make-up can impact the susceptibility of a host to infectious disease and the magnitude of the host immune response. In turn, when considering multiple host species and their interactions at the community level, these features serve as important indicators of ecosystem health, in the context of zoonotic potential and disease transmissibility.

Disease spillover has become an increasing concern as human populations continue to increase in density and encroach on wild habitats.

Incidents of disease spillover have been increasingly documented throughout the last century. Spillover of zoonotic pathogens has been attributed to a number of anthropogenic activities, including (but not limited to) mass deforestation, habitat encroachment, mass production of livestock, and urbanization. In addition to bringing humans and domestic animals into closer and more frequent contact, land-use shifts from natural to urban provide increased breeding grounds for many arthropod vectors of disease. Nutrient removal by slash and burn practices, nitrification of run-off waters, and husbandry practices that confine multiple species into single factories all contribute to survival challenges for wildlife, and increased stress responses for both wild and domestic animals (Ditchkoff *et al.*, 2006; Winfree, 2009; Rolland *et al.*, 2012; Strasser *et al.*, 2013; Newbolt *et al.*, 2017).

These well-known anthropogenic effects describe only some of the challenges facing wildlife health. Dysbiosis – the disruption of the microbiome – is a potentially significant symptom of anthropogenic disturbance. Given what we now know about the evolutionary links between microbes and their hosts, the impact microbes can have on the stress response and other endocrinological functions of animals, and the effect of increased stress on fitness and survival, it is clear that the microbiome can have profound effects on the health and survival of not only individuals, but of long-term populations. Furthermore, dysbiosis can lead to increased pathogenicity and, in the case of viral infections, increased viral shedding. Viral shedding is one direct mechanism of viral transmission and has been strongly implicated in zoonotic transmission between many animals (both wild and domestic) and humans.

Zoonoses can affect humans both directly (e.g. Marburg, SARS) and indirectly (e.g. EEE, avian influenza, mycoplasmas affecting livestock and domestic animals). In addition to detrimental effects on human and domestic animal health, zoonotic diseases can have indirect devastating effects on the wild population from which they originate. For example, mass cullings of wild animals, such as bats and primates, sometimes occur when local communities deem these populations too great of a threat [see Pooley *et al.* (2015) and references therein].

Refined spatio-temporal variation data can improve accuracy of epidemic models (e.g. Olivero *et al.*, 2017). Similarly, the inclusion of microbial symbiont data has the power to inform epidemiological models and aid in outbreak predictions, mainly by developing a better understanding of when wildlife populations are healthy, when they are not, and how dysbiosis can amplify zoonotic disease transmission. Microbial data can help identify mitigating risk factors and potentially improve development of prophylactics and treatments. For example, studies of the skin microbiome in North American *Myotis* bats have revealed antifungal properties of naturally occurring bacteria that may protect populations from the deadly fungus, *Pseudogymnoascus destructans* (the cause of the devastating White-nose Syndrome, which has killed nearly six million bats since its emergence in North America in 2006; Lemieux-Labonte *et al.*, 2017). Similarly, novel probiotic treatments have been proposed to help amphibian populations suffering from a different devastating fungal infection, caused by the chytrid fungus *Batrachochytrium dendrobatidis* (Bletz *et al.*, 2013). Such observations and potential solutions to wildlife diseases (which have most likely been caused by anthropogenic disruptions) highlight the importance of characterizing the microbial make-up of natural populations. In essence, understanding which symbionts are 'normally' present can allow anomalies to be detected and assessed in a larger epidemiological framework.

We conclude this chapter with a quote from the vision statement of the One Health Initiative – a worldwide consortium and self-proclaimed 'strategy for expanding interdisciplinary collaborations and communications in all aspects of health care for humans, animals and the environment':

> Recognizing that human health (including mental health via the human–animal bond phenomenon), animal health, and ecosystem health are inextricably linked, One Health seeks to promote, improve, and defend the health and well-being of all species by enhancing cooperation and collaboration between physicians, veterinarians, other scientific health and environmental professionals and by promoting strengths in leadership and management to achieve these goals.
>
> One Health Initiative, Vision Statement 2018, www.onehealth.com/mission

The relationships between animal health, ecosystem health, and human health are inextricably intertwined. One component is not immune to the detrimental effects experienced by another, and the health of the system as a whole depends on the health of each component. Defining and monitoring wildlife health is a challenging area of research, and one that is benefiting significantly from the improved accessibility of high-throughput sequencing and 'omics' technologies. The establishment of baseline data for wildlife communities using these approaches will help establish effective definitions of 'health' for animals. This in turn will lead to improved characterization of ecosystem health, and will allow researchers to define and monitor communities. This is particularly important at the human–wildlife interface, where microbial associations need to be rapidly identified as normal trends or potential sources of disease either for wildlife or humans. Refined monitoring of local communities will provide important context for zoonotic outbreaks and pathogen spillover events, and allow researchers to improve both predictions and preventative approaches to disease. This comprehensive approach ultimately serves to benefit both humans and wildlife.

References

Amann, R.I., Ludwig, W., and Schleifer, K.H. (1995). Phylogenetic identification and in situ detection of individual microbial cells without cultivation. Microbiol. Rev. *59*, 143–169.

Ariefdjohan, M.W., Savaiano, D.A., and Nakatsu, C.H. (2010). Comparison of DNA extraction kits for PCR-DGGE analysis of human intestinal microbial communities from fecal specimens. Nutr. J. *9*, 23. https://doi.org/10.1186/1475-2891-9-23

Bahrndorff, S., Alemu, T., Alemneh, T., and Lund Nielsen, J. (2016). The microbiome of animals: implications for conservation biology. Int. J. Genomics *2016*, 5304028. https://doi.org/10.1155/2016/5304028

Baker, M.R., Gobush, K.S., & Vynne, C.H. (2013). Review of factors influencing stress hormones in fish and wildlife. J. Nat. Conserv. *21*, 309–318.

Barelli, C., Albanese, D., Donati, C., Pindo, M., Dallago, C., Rovero, F., Cavalieri, D., Tuohy, K.M., Hauffe, H.C., and De Filippo, C. (2015). Habitat fragmentation is associated to gut microbiota diversity of an endangered primate: implications for conservation. Sci. Rep. *5*, 14862. https://doi.org/10.1038/srep14862

Besedovsky, H.O., and del Rey, A. (1996). Immune-neuro-endocrine interactions: facts and hypotheses. Endocr. Rev. *17*, 64–102. https://doi.org/10.1210/edrv-17-1-64

Bestion, E., Jacob, S., Zinger, L., Di Gesu, L., Richard, M., White, J., and Cote, J. (2017). Climate warming reduces gut microbiota diversity in a vertebrate ectotherm. Nat. Ecol. Evol. *1*, 161. https://doi.org/10.1038/s41559-017-0161

Bletz, M.C., Loudon, A.H., Becker, M.H., Bell, S.C., Woodhams, D.C., Minbiole, K.P., and Harris, R.N. (2013). Mitigating amphibian chytridiomycosis with bioaugmentation: characteristics of effective probiotics and strategies for their selection and use. Ecol. Lett. *16*, 807–820. https://doi.org/10.1111/ele.12099

Boonstra, R. (2005). Equipped for life: The adaptive role of stress axis in male mammals. J. Mammal. *86*, 236–247.
Boonstra, E., de Kleijn, R., Colzato, L.S., Alkemade, A., Forstmann, B.U., and Nieuwenhuis, S. (2015). Neurotransmitters as food supplements: the effects of GABA on brain and behavior. Front. Psychol. *6*, 1520. https://doi.org/10.3389/fpsyg.2015.01520
Bravo, J.A., Forsythe, P., Chew, M.V., Escaravage, E., Savignac, H.M., Dinan, T.G., Bienenstock, J., and Cryan, J.F. (2011). Ingestion of Lactobacillus strain regulates emotional behavior and central GABA receptor expression in a mouse via the vagus nerve. Proc. Natl. Acad. Sci. U.S.A. *108*, 16050–16055. https://doi.org/10.1073/pnas.1102999108
Ceballos, G., Ehrlich, P.R., Barnosky, A.D., García, A., Pringle, R.M., and Palmer, T.M. (2015). Accelerated modern human-induced species losses: Entering the sixth mass extinction. Sci. Adv. *1*, e1400253. https://doi.org/10.1126/sciadv.1400253
Charmandari, E., Tsigos, C., and Chrousos, G. (2005). Endocrinology of the stress response. Annu. Rev. Physiol. *67*, 259–284. https://doi.org/10.1146/annurev.physiol.67.040403.120816
Chen, C., Song, X., Wei, W., Zhong, H., Dai, J., Lan, Z., Li, F., Yu, X., Feng, Q., Wang, Z., *et al.* (2017). The microbiota continuum along the female reproductive tract and its relation to uterine-related diseases. Nat. Commun. *8*, 875. https://doi.org/10.1038/s41467-017-00901-0
Costa, S., Lopes, I., Proença, D.N., Ribeiro, R., and Morais, P.V. (2016). Diversity of cutaneous microbiome of *Pelophylax perezi* populations inhabiting different environments. Sci. Total Environ. *572*, 995–1004.
DeCandia, A.L., Dobson, A.P., and vonHoldt, B.M. (2018). Toward an integrative molecular approach to wildlife disease. Conserv. Biol. *32*, 798–807. https://doi.org/10.1111/cobi.13083
Diaz Heijtz, R., Wang, S., Anuar, F., Qian, Y., Björkholm, B., Samuelsson, A., Hibberd, M.L., Forssberg, H., and Pettersson, S. (2011). Normal gut microbiota modulates brain development and behavior. Proc. Natl. Acad. Sci. U.S.A. *108*, 3047–3052. https://doi.org/10.1073/pnas.1010529108
Ditchkoff, S.S., Saalfeld, S.T., and Gibson, C.J. (2006). Animal behavior in urban ecosystems: modifications due to human-induced stress. Urban Ecosyst. *9*, 5–12.
Ehrlich, P.R., and Ehrlich, A.H. (2013). Can a collapse of global civilization be avoided? Proc. Biol. Sci. *280*, 20122845. https://doi.org/10.1098/rspb.2012.2845
Ezenwa, V.O., Gerardo, N.M., Inouye, D.W., Medina, M., and Xavier, J.B. (2012). Microbiology. Animal behavior and the microbiome. Science *338*, 198–199. https://doi.org/10.1126/science.1227412
Foster, J.A., and McVey Neufeld, K.A. (2013). Gut-brain axis: how the microbiome influences anxiety and depression. Trends Neurosci. *36*, 305–312. https://doi.org/10.1016/j.tins.2013.01.005
Freeman, E.W., Meyer, J.M., Adendorff, J., Schulte, B.A., and Santymire, R.M. (2014a). Scraping behavior of black rhinoceros is related to age and fecal gonadal metabolite concentrations. J. Mammal. *95*, 340–348.
Freeman, E.W., Meyer, J.M., Bird, J., Adendorff, J., Schulte, B.A., and Santymire, R.M. (2014b). Impacts of environmental pressures on the reproductive physiology of subpopulations of black rhinoceros (*Diceros bicornis bicornis*) in Addo Elephant National Park, South Africa. Conserv. Physiol. *2*, cot034. https://doi.org/10.1093/conphys/cot034
Fritz, B.A., Thomas, C.B., and Yuill, T.M. (1992). Serological and microbial survey of *Mycoplasma galliscepticum* in wild turkeys (*Meleagris gallopavo*) from six western states. J. Wildl. Dis. *28*, 10–20. https://doi.org/10.7589/0090-3558-28.1.10
Goymann, W. (2012). On the use of non-invasive hormone research in uncontrolled, natural environments: the problem with sex, diet, metabolic rate and the individual. Meth. Ecol. Evol. *3*, 757–765.
Harper, J.M., and Austad, S.N. (2000). Fecal glucocorticoids: a noninvasive method of measuring adrenal activity in wild and captive rodents. Physiol. Biochem. Zool. 73, 12–22.
Harper, J.M., and Austad, S.N. (2004). Fecal corticosteroid levels in free-living populations of deer mice (*Peromyscus maniculatus*) and southern red-backed voles (*Clethrionomys gapperi*). Amer. Midl. Nat. *152*, 400–409.
Hing, S., Narayan, E.J., Thompson, R., Andrew, C., and Godfrey, S.S. (2016). The relationship between physiological stress and wildlife disease: consequences for health and conservation. Wildlife Res. *43*, 51–60.
Huber, M., Knottnerus, J.A., Green, L., van der Horst, H., Jadad, A.R., Kromhout, D., Leonard, B., Lorig, K., Loureiro, M.I., van der Meer, J.W.M., *et al.* (2011). How should we define health?. Br. Med. J. 343, d4163.
Kakumanu, M.L., Reeves, A.M., Anderson, T.D., Rodrigues, R.R., and Williams, M.A. (2016). Honey bee gut microbiome is altered by in-hive pesticide exposures. Front. Microbiol. 7, 1255. https://doi.org/10.3389/fmicb.2016.01255

Kohl, K.D. (2012). Diversity and function of the avian gut microbiota. J. Comp. Physiol. B, Biochem. Syst. Environ. Physiol. *182*, 591–602. https://doi.org/10.1007/s00360-012-0645-z

Kumar, A., Roberts, D., Wood, K.E., Light, B., Parrillo, J.E., Sharma, S., Suppes, R., Feinstein, D., Zanotti, S., Taiberg, L., *et al.* (2006). Duration of hypotension before initiation of effective antimicrobial therapy is the critical determinant of survival in human septic shock. Crit. Care Med. *34*, 1589–1596. https://doi.org/10.1097/01.CCM.0000217961.75225.E9

Langille, M.G., Zaneveld, J., Caporaso, J.G., McDonald, D., Knights, D., Reyes, J.A., Clemente, J.C., Burkepile, D.E., Vega Thurber, R.L., Knight, R., *et al.* (2013). Predictive functional profiling of microbial communities using 16S rRNA marker gene sequences. Nat. Biotechnol. *31*, 814–821. https://doi.org/10.1038/nbt.2676

Laupland, K.B., and Valiquette, L. (2013). The changing culture of the microbiology laboratory. Can. J. Infect. Dis. Med. Microbiol. *24*, 125–128.

Lemieux-Labonté, V., Simard, A., Willis, C.K.R., and Lapointe, F.J. (2017). Enrichment of beneficial bacteria in the skin microbiota of bats persisting with white-nose syndrome. Microbiome *5*, 115. https://doi.org/10.1186/s40168-017-0334-y

Lombardo, M.P. (1998). On the evolution of sexually transmitted diseases in birds. J. Avian Biol. *29*, 314–321.

Lombardo, M.P., and Thorpe, P.A. (2000). Microbes in tree swallow semen. J. Wildl. Dis. *36*, 460–468. https://doi.org/10.7589/0090-3558-36.3.460

Lombardo, M.P., Thorpe, P.A., Cichewicz, R., Henshaw, M., Millard, C., Steen, C., and Zeller, T.K. (1996). Communities of cloacal bacteria in tree swallow families. Condor *98*, 167–172.

Lyte, M. (2013). Microbial endocrinology in the microbiome-gut-brain axis: how bacterial production and utilization of neurochemicals influence behavior. PLOS Pathog. *9*, e1003726. https://doi.org/10.1371/journal.ppat.1003726

Mace, G.M., Norris, K., and Fitter, A.H. (2012). Biodiversity and ecosystem services: a multilayered relationship. Trends Ecol. Evol. *27*, 19–26. https://doi.org/10.1016/j.tree.2011.08.006

Martin, P., Bateson, P.P.G., and Bateson, P. (2010). Measuring Behavior: An Introductory Guide. (Cambridge University Press, Cambridge).

Martín-Vivaldi, M., Ruiz-Rodríguez, M., Soler, J.J., Manuel Peralta-Sánchez, J., Méndez, M., Valdivia, E., Martín-Platero, A.M., and Martínez-Bueno, M. (2009). Seasonal, sexual and developmental differences in hoopoe Upupa epops preen gland morphology and secretions: evidence for a role of bacteria. J. Avian Biol. *40*, 191–205.

McEwen, B.S., and Wingfield, J.C. (2003). The concept of allostasis in biology and biomedicine. Horm. Behav. *43*, 2–15.

Míguez, I., and Aldegunde, M.A. (1990). Effect of gamma-aminobutyric acid on corticosterone secretion: involvement of the noradrenergic system. Life Sci. *46*, 875–880.

Millspaugh, J.J., and Washburn, B.E. (2004). Use of fecal glucocorticoid metabolite measures in conservation biology research: considerations for application and interpretation. Gen. Comp. Endocrinol. *138*, 189–199. https://doi.org/10.1016/j.ygcen.2004.07.002

Monfort, S.L., Holt, W.V., Pickard, A.R., Rodger, J.C., and Wildt, D.E. (2003). Non-invasive endocrine measures of reproduction and stress in wild populations. Conservation Biology Series *8*, 147–165. (Cambrige University Press, Cambridge)

Moreno, J., Briones, V., Merino, S., Ballesteros, C., Sanz, J.J., and Tomás, G. (2003). Beneficial effects of cloacal bacteria on growth and fledging size in nestling pied flycatchers (*Ficedula hypoleuca*) in Spain. Auk *120*, 784.

Moreno-Rueda, G. (2010). Uropygial gland size correlates with feather holes, body condition and wingbar size in the house sparrow *Passer domesticus*. J. Avian Biol. *41*, 229–236.

Moreno-Rueda, G. (2011). House Sparrows *Passer domesticus* with larger uropygial glands show reduced feather wear. Ibis *153*, 195–198.

Moss, A., Jensen, E., and Gusset, M. (2015). Evaluating the contribution of zoos and aquariums to Aichi Biodiversity Target 1. Conserv. Biol. *29*, 537–544. https://doi.org/10.1111/cobi.12383

Möstl, E., and Palme, R. (2002). Hormones as indicators of stress. Domest. Anim. Endocrinol. *23*, 67–74.

Neuman, H., Debelius, J.W., Knight, R., and Koren, O. (2015). Microbial endocrinology: the interplay between the microbiota and the endocrine system. FEMS Microbiol. Rev. *39*, 509–521. https://doi.org/10.1093/femsre/fuu010

Newbolt, C.H., Acker, P.K., Neuman, T.J., Hoffman, S.I., Ditchkoff, S.S., and Steury, T.D. (2017). Factors influencing reproductive success in male white-tailed deer. J. Wildlife Manag. *81*, 206–217.

Nicholson, J.K., Holmes, E., Kinross, J., Burcelin, R., Gibson, G., Jia, W., and Pettersson, S. (2012). Host-gut microbiota metabolic interactions. Science *336*, 1262–1267. https://doi.org/10.1126/science.1223813

Olivero, J., Fa, J.E., Real, R., Márquez, A.L., Farfán, M.A., Vargas, J.M., Gaveau, D., Salim, M.A., Park, D., Suter, J., *et al.* (2017). Recent loss of closed forests is associated with Ebola virus disease outbreaks. Sci. Rep. *7*, 14291. https://doi.org/10.1038/s41598-017-14727-9

Padgett, D.A., and Glaser, R. (2003). How stress influences the immune response. Trends Immunol. *24*, 444–448.

Perek, M., Elian, M., and Heller, E.D. (1969). Bacterial flora of semen and contamination of the reproductive organs of the hen following artificial insemination. Res. Vet. Sci. *10*, 127–132.

Poessel, S.A., Biggins, D.E., Santymire, R.M., Livieri, T.M., Crooks, K.R., and Angeloni, L. (2011). Environmental enrichment affects adrenocortical stress responses in the endangered black-footed ferret. Gen. Comp. Endocrinol. *172*, 526–533. https://doi.org/10.1016/j.ygcen.2011.04.018

Poiani, A., and Wilks, C. (2000). Sexually transmitted diseases: a possible cost of promiscuity in birds?' Auk *117*, 1061.

Pooley, S., Fa, J.E., and Nasi, R. (2015). No conservation silver lining to Ebola. Conserv. Biol. *29*, 965–967. https://doi.org/10.1111/cobi.12454

Reeder, D.M., and Kramer, K.M. (2005). Stress in free-ranging mammals: Integrating physiology, ecology, and natural history. J. Mammal. *86*, 225–235.

Rolland, R.M., Parks, S.E., Hunt, K.E., Castellote, M., Corkeron, P.J., Nowacek, D.P., Wasser, S.K., and Kraus, S.D. (2012). Evidence that ship noise increases stress in right whales. Proc. Biol. Sci. *279*, 2363–2368. https://doi.org/10.1098/rspb.2011.2429

Romero, L.M. (2004). Physiological stress in ecology: lessons from biomedical research. Trends Ecol. Evol. *19*, 249–255.

Rothschild, D., Weissbrod, O., Barkan, E., Kurilshikov, A., Korem, T., Zeevi, D., Costea, P.I., Godneva, A., Kalka, I.N., Bar, N., *et al.* (2018). Environment dominates over host genetics in shaping human gut microbiota. Nature *555*, 210–215. https://doi.org/10.1038/nature25973

Ruiz-Rodríguez, M., Soler, J.J., Martín-Vivaldi, M., Martín-Platero, A.M., Méndez, M., Peralta-Sánchez, J.M., Ananou, S., Valdivia, E., and Martínez-Bueno, M. (2014). Environmental factors shape the community of symbionts in the hoopoe uropygial gland more than genetic factors. Appl. Environ. Microbiol. *80*, 6714–6723. https://doi.org/10.1128/AEM.02242-14

Ryser-Degiorgis, M.P. (2013). Wildlife health investigations: needs, challenges and recommendations. BMC Vet. Res. *9*, 223. https://doi.org/10.1186/1746-6148-9-223

Santymire, R.M., and Armstrong, D.M. (2010). Development of a field-friendly technique for fecal steroid extraction and storage using the African wild dog (*Lycaon pictus*). Zoo Biol. *29*, 289–302. https://doi.org/10.1002/zoo.20248

Sapolsky, R.M. (2001). Physiological and pathophysiological implications of social stress in mammals. In Handbook of Physiology IV, McEwen, B.S., and Goodman, H.M., eds. (Oxford University Press, New York), pp. 517–532.

Sender, R., Fuchs, S., and Milo, R. (2016). Revised estimates for the number of human and bacteria cells in the body. PLOS Biol. *14*, e1002533. https://doi.org/10.1371/journal.pbio.1002533

Soler, J.J., Martín-Vivaldi, M., Ruiz-Rodríguez, M., Valdivia, E., Martín-Platero, A.M., Martínez-Bueno, M., Peralta-Sánchez, J.M., and Méndez, M. (2008). Symbiotic association between hoopoes and antibiotic-producing bacteria that live in their uropygial gland. Funct. Ecol. *22*, 864–871.

Song, S.J., Amir, A., Metcalf, J.L., Amato, K.R., Xu, Z.Z., Humphrey, G., and Knight, R. (2016). preservation methods differ in fecal microbiome stability, affecting suitability for field studies. mSystems *1*, e00021-16. https://doi.org/10.1128/mSystems.00021-16

Stephen, C. (2014). Toward a modernized definition of wildlife health. J. Wildl. Dis. *50*, 427–430. https://doi.org/10.7589/2013-11-305

Stipkovits, L., Varga, Z., Czifra, G., and Dobos-Kovacs, M. (1986). Occurrence of mycoplasmas in geese affected with inflammation of the cloaca and phallus. Avian Pathol. *15*, 289–299.

Strasser, E.H., Heath, J.A., and Thompson, D. (2013). Reproductive failure of a human-tolerant species, the American kestrel, is associated with stress and human disturbance. J. Appl. Ecol. *50*, 912–919.

Sudo, N., Chida, Y., Aiba, Y., Sonoda, J., Oyama, N., Yu, X.N., Kuno, C., and Koga, Y. (2004). Postnatal microbial colonization programs the hypothalamic-pituitary-adrenal system for stress response in mice. J. Physiol. *558*, 263–275.

Tarlow, E.M., and Blumstein, D.T. (2007). Evaluating methods to quantify anthropogenic stressors on wild animals. Appl. Anim. Behav. Sci. *102*, 429–451.

Thompson, L.R., Sanders, J.G., McDonald, D., Amir, A., Ladau, J., Locey, K.J., Prill, R.J., Tripathi, A., Gibbons, S.M., Ackermann, G., *et al.* (2017). A communal catalogue reveals Earth's multiscale microbial diversity. Nature *551*, 457–463.

Touma, C., and Palme, R. (2005). Measuring fecal glucocorticoid metabolites in mammals and birds: the importance of validation. Ann. N. Y. Acad. Sci. *1046*, 54–74.

Wasser, S.K., Hunt, K.E., Brown, J.L., Cooper, K., Crockett, C.M., Bechert, U., Millspaugh, J.J., Larson, S., and Monfort, S.L. (2000). A generalized fecal glucocorticoid assay for use in a diverse array of nondomestic mammalian and avian species. Gen. Comp. Endocrinol. *120*, 260–275. https://doi.org/10.1006/gcen.2000.7557

Welsh, T.H.J., Kemper-Green, C.N., and Livingston, K.N. (1999). Stress and reproduction. In Encyclopedia of Reproduction, Neill, J.D., and Knobil, E., eds. (Academic Press, New York), pp. 662–674.

Westneat, D.F., and Rambo, T.B. (2000). Copulation exposes female Red-winged Blackbirds to bacteria in male semen. J. Avian Biol. *31*, 1–7.

White, J., Mirleau, P., Danchin, E., Mulard, H., Hatch, S.A., Heeb, P., and Wagner, R.H. (2010). Sexually transmitted bacteria affect female cloacal assemblages in a wild bird. Ecol. Lett. *13*, 1515–1524. https://doi.org/10.1111/j.1461-0248.2010.01542.x

Williams, M.R., and Gallo, R.L. (2017). Evidence that human skin microbiome dysbiosis promotes atopic dermatitis. J. Invest. Dermatol. *137*, 2460–2461.

Winfree, R., Aguilar, R., Vázquez, D.P., LeBuhn, G., and Aizen, M.A. (2009). A meta-analysis of bees' responses to anthropogenic disturbance. Ecology *90*, 2068–2076.

Wingfield, J.C. (2005). The concept of allostasis: Coping with a capricious environment. J. Mammal. *86*, 248–254.

Wingfield, J.C., and Romero, L.M. (2001). Adrenocortical responses to stress and their modulation in free-living vertebrates. In Handbook of Physiology: A Critical, Comprehensive Presentation of Physiological Knowledge and Concepts, IV, McEwen, B.S., and Goodman, H.M., eds. (Oxford University Press, Oxford), pp. 211–234.

Index

www.ingramcontent.com/pod-product-compliance
Lightning Source LLC
LaVergne TN
LVHW051050210926
848486LV00001B/64
* 9 7 8 1 9 1 2 5 3 0 0 2 1 *